Saint Anthony's Fire

LSD from the Air Force to the Summer of Love

A MEMOIR

Saint Anthony's Fire

LSD from the Air Force to the Summer of Love

A MEMOIR

Michael Stanis

Grizzly Peak Press
Kensington ~ CA

For information contact:
Grizzly Peak Press
350 Berkeley Park Blvd.
Kensington, CA 94707
www.grizzlypeakpress.com

Saint Anthony's Fire
is published by Daniel N. David
and is distributed by
Grizzly Peak Press.

Design, layout and typesetting by
Liquid Pictures
www.liquidpictures.com

ISBN Number: 978-1-950393-91-6
Library of Congress Number: 2019934981

Printed in the United States of America

Saint Anthony's Fire

LSD from the Air Force to the Summer of Love

A MEMOIR

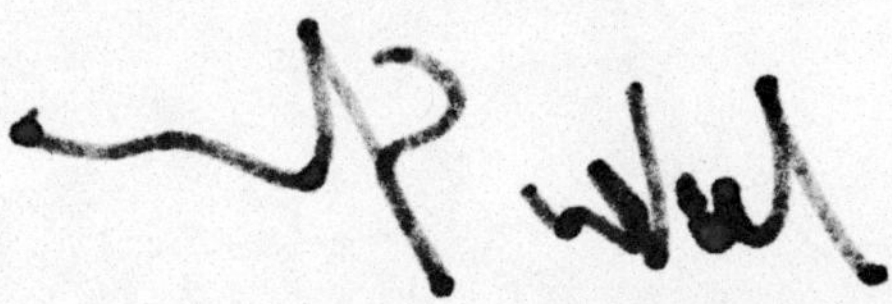

CONTENTS

Saint Anthony's Fire

In medieval times, villages in Northern Europe that depended on rye cultivation would sometimes experience a mysterious outbreak of mass hallucinations, one of which was that objects would seem to glow, even be on fire.

If the weather was too wet around the time of the harvest, a fungus, ergot, that's used to treat migraine now, approximately the size of rye grain would grow on the plants and end up being ground up with these grains and made into bread.

St. Anthony was associated with the effort to ameliorate the peasants' distress. It was noticed that those who did not eat rye, and ate wheat bread instead, were unaffected. St.Anthony opened dining rooms to serve untainted bread and today, dining rooms bearing his name serve meals and provide social services in many urban areas.

In 1943, the year of my birth, Albert Hoffman, a Swiss chemist, was working on the synthesis of compounds from ergotamine, derived from the fungus. Lysergic Acid Diethylamide was one of these, and as he was working with the 25th iteration of the molecule he got some on his hands. As he bicycled home from the lab, he noticed strange feelings, such as his legs seeming to stretch out.

He resumed work on this extraordinary substance. A very small dose could produce unusual effects.

Hoffman had discovered LSD-25, which was eventually manufactured by Sandoz Pharmaceuticals.

During the fifties, this remarkable drug came to the attention of our military and intelligence services, who needed to know how it could be used, partially to be one step ahead of our geopolitical rivals.

To this end, Allen Dulles authorized experiments to be conducted by Dr. Ewen Cameron at a mental hospital in Montreal, by

Dr. Sidney Gottlieb and by the U.S.army at Ft. Detrick, Maryland. In one instance, Major Frank Olson ended up dead on the sidewalk outside a hotel, having fallen from a window under murky circumstances, giving rise to the opinion that LSD made you think you could fly.

In 1961, at Lackland air Force Base, the compound was tested on volunteers in basic training.

So it was that in St.Anthony's city, Albert Hoffman's discovery was administered to an airman born in the year of its discovery.

Military Roots

~

Today, universal military service, the draft, is a memory in these United States. In 1961, for those of us born male in 1943, it was a looming reality. Having registered, and having been classified 4A, my choices were to wait to be drafted, or to preemptively choose a branch of the service and enlist.

My father, Paul Stanis, who was a professional musician in New Jersey, when the U.S.entered the Second World War, was drafted in 1942 and assigned to the 509th Military Band at Jefferson Barracks, near St.Louis, Missouri. He worked as an arranger/ pianist and played glockenspiel in the marching band. The arrangement of W.C.Handy's St.Louis Blues March, featured on Glenn Miller's Air Force Band album was developed there.

Paul married my mother, Jeanette, the great granddaughter of a Chickasaw woman who married a Welshman and left the tribe about the time of removal, about 1837, moving north to Indiana to live as a white family, and I was born at St. Mary's Hospital in February of 1943.

My great-grandfather was a teenage drummer in the 17th Illinois Volunteers during the

Civil War, in the Union Army, and I had an uncle, my father's brother Walter, who had served during the Invasion of Italy and the Battle of the Bulge. My maternal uncle, James, was a Navy CPO who was in the Battle of the Coral Sea.

I chose the Air Force, the successor to my father's Army Air Corps.

I was informed in a letter from my recruiter that I was eligible to join and on May 30th,

1961, went to the Induction Center for processing. After some paperwork, we had a Physical, undressing to our underwear, and, as they say, being poked and prodded. More paperwork. I had disclosed that an 11 year old neighbor had had his way with me when I was four, but the Psych judged that I was eligible to join.

Finally, about thirty of us were ready to be sworn in. There was an air of enthusiasm to be putting ourselves at the service of our new Commander-in-Chief, John F. Kennedy. We swore to protect and defend the Constitution of the United States against all enemies, foreign and domestic, and to faithfully execute the orders of our commander in chief and his chain of command.

We were cautioned that if anything happened to Kennedy, our allegiance would be to his successor.

Next, we were put on buses to Union Station, where we boarded for our basic training at Lackland Air Force Base in San Antonio, Texas. We were figuratively a motley group, with our varied civilian clothing. We were mostly white, with a few blacks and latinos, and at least one Chickasaw Indian passing for white.

Good spirits prevailed in our car. Someone had a guitar and I played and sang "Malaguena Salerosa"-"Que Bonitos Ojos Tienes" which I had learned from a Theodore Bikel songbook, surprising the Mexican- Americans on board.

We arrived at Lackland the next afternoon, where we were fed at a civilian-style Cafeteria, and marched to our barracks. We were to be Basic Training Flight 866. Lights out at 8:50 PM. Next day at 4:45 AM, the real work would begin.

Ten minutes, S,S, and S, as they say, get dressed and fall out in front of the barracks, in line for roll call. We're wearing reflective harnesses so drivers can see us as we march in the pre-dawn darkness. We shamble to the mess hall. You couldn't call it marching at this point. We are a sight for derision because of our civvies- next week we would be able to laugh at the new crop of "rainbows".

Food was served out on metal trays, reconstituted dehydrated scrambled eggs, bacon or some other meat, pancakes or biscuits, juice, and coffee. A short time was allotted, then we mustered outside for our next duty, which was to be shorn of our civilian hairstyles. Some guys had prepared-they had gotten crew cuts or flattops, but were subjected to the same #1 buzzcuts endured by the most longhaired of us. The blacks were in another line, with barbers more familiar with the exigencies of their hair. As it turns out, those haircuts looked better than those of the whites.

Now we had "uniform" heads, time for uniform clothing.

coffee. A short time was allotted, then we were directed outside for our next duty, which was to be shorn of our civilian haircuts. Some guys had prepared—they had gotten crew cuts or flattops, but were subjected to the same 45 seconds endured by the most long-haired of us. The blacks were in another line with barbers more familiar with the exigencies of their hair. As it turned out, those haircuts looked better than those of the whites.

Now we had "uniform" heads. Time for uniform clothing.

The Green Monster

There was a sprawling processing center called "the green monster", where, among other things, we were fitted for our uniforms. We were measured. Then we were issued a duffel bag then proceeded to another line where we tried on the first of our uniforms to confirm the measurements. They seemed too big, but we were assured we would gain weight during our training. We began filling the duffel- a dress uniform, wool pants and a jacket which they called a "blouse", to a certain amount of snickering. Summer uniform, khaki pants, blue shirts, tan shirts, a short khaki jacket, which we would find out we could never wear since it turned out it was reserved for the CIA guys, underwear, socks and two caps, one to go with the wool uniform and one for the khaki, named after a disparaging term for a woman, or for an anatomical feature of that woman.

After all, we found out we would seldom have a chance to wear any of it, aside from the underwear. Our daily working clothes were "fatigues", delivered as needed by the base laundry. Our shoes were black Oxfords and "Brogans", a ¾ boot, laced, also black.

This same busy day we would learn how to stow our clothing in our footlockers. Everything went in there, the folded duffel, all our uniforms, our properly folded underwear, all our orders, our oxfords, but not our brogans.

We learned to polish our shoes, practicing the art of spit-shining. Our next task was making up our double-decker bunks. We learned hospital corners and how to make the blanket tight enough to bounce a quarter on.

"Road Guards Out!"

~

The next morning, we looked like airmen, except for the marching, which was to be the agenda for the day. First the commands; attention, right face, left face, about face, parade rest, at ease, The marching commands; forward march, to the rear march, column right and column left march, right and left oblique march, right flank and left flank march, column halt.

We marched all over the base and started to get good at it, but I was a little reactive,and was not so smooth in my rank, so I got assigned as a road guard. At the command "road guards out!", the two of us at the front of the column would sprint out to the intersection and hold out our stop sign and blow the whistle. As the column was halfway through the intersection the rear road guards would take our places and we would reenter the column. I actually liked this since I had something to think about while marching.

Later in the afternoon, we broke into smaller groups for close order drill, groups of four to eight. It was a little like square dancing, and was fun once you got it. Unfortunately, a few of us nerds were having trouble with it, and the sergeant had the rest go in, and several of us had remained with a two-striper. The sergeant had taken him aside and said something. After he was gone, the airman produced a joint and had us each take a couple puffs. He put us through our paces, we found our groove and successfully fulfilled our close order drill training.

After we rejoined our flight, we had a leadership lecture. I was still feeling the pot and everything the captain was saying seemed like gobbledygook. We were asked to nominate and vote on squad leaders. I just went along with the consensus.

After dinner, we were mostly done with our duties and it was announced that the smoking lamp was lit. The rules were explained and we were allowed our first smokes in the barracks.

The next day, we filled out forms in a pleasant building, I think it was a chapel hall or annex. One question was whether we were conscientious objectors. I had given this matter some thought in high school, so I ticked the box. This answer was not acceptable, since it was reserved for members of a particular religion officially holding this tenet. As to the question of religious affiliation, I checked the box for "Catholic", which would appear on my dog tags. I remembered how we kids had grown up in St. Louis, being brought to a variety of churches by our mother. First a Cherokee Church held in a storefront, then a Baptist Church, which, in a picture in the Sunday School, portrayed Jesus as a blond Confederate Cavalry officer, though I may have had that impression from hearing the place name Calvary.

We were questioned why our father wasn't there- of course being a musician he worked nights and slept in, sunday mornings.The people there drew my mother's ire when they described us kids as "wild Indians". We never went back.

Our next church was a Church of the Nazarene. We were welcome there and that is where I learned the major stories of the Old Testament. Though it was called the Church of the Nazarene, it may as well have been a Jewish Synagogue, since the sermons were almost exclusively drawn from the Old Testament.

When we moved to a new house in the suburbs, Northwoods Mo.,we joined St. Andrew's Episcopal Church,where I learned about liberal Christianity, which accepted evolution and believed in social justice. My sisters and I were in the choir, where I fell in love with Miss Gallatin, the director. I later became an altar boy. Before the service the reverend and I would say prayers and have a little nip of the sacramental sherry in the vestry. The devices to light the altar candles were about ten feet long, and sometimes I had trouble lighting the wicks of the tallest candles. The reverend was later relieved, and communion became somewhat less jolly.

A friend from down the street, Charlie, talked me into going to Catechism classes at The Church of the Assumption, to have reinforcements against advances by the Priest there. I'm not sure why he wanted to convert, but since Catholicism was my paternal grandfather's religion, I accompanied him in the effort. We got confirmed, and became friendlier with the Italian girls in our neighborhood.

Someone came out and asked for volunteers for a study for a medication to help mentally ill people. Thinking I might get out of some marching, I agreed. A small number of us signed a waiver for the purpose, giving up our right to sue the Air Force or the pharmaceutical company. It happened that the "medication" was LSD, and the testing was for other purposes than helping people with mental illness. We were not told when the testing would begin.

Chain of Command

The recruit is at the bottom of the chain of command. From the top, here is my understanding of it:

At the top is the President of the United States; he with the help of the Secretary of Defense, oversees the Joint Chiefs of Staff, and the Secretaries of the various services.

The chief of staff of the Air Force supervises the heads of the various commands, Strategic Air Command, for example and, in my case, the head of the Air Training Command.

Under him, the commanders of the training bases, in this case Lackland Air Force Base, home of all basic training in the AF.

Several squadrons answer to the base commander, and various companies, typically commanded by a captain, watch over a group of flights, each in their own barracks. Within the barracks, there are two floors, each with two rows of bunks. The Airmen in each row make up a squad. (If I remember right, each floor is a platoon).

It was the squad leaders who we elected the day before.

One day, the agenda was future work assignments. These were the jobs we would perform and the specialties that would define us through our enlistments. The two most numerous specialties were Air Police, now called Security Services, the AF equivalent of MP, and the Medics, encompassing all the hospital jobs, and the Preventive Medicine specialty concerned with Public Health issues and Industrial safety, as well as issues of nuclear, chemical, and biological warfare.

I had tested as being qualified for three jobs; Air Traffic Control, Meteorology and Preventive Med. Before we found out our assignments, we had a round of immunizations.

We went from one long line to another, sleeves rolled up, and were given four or five shots. The implement used did not involve needles. The injector fired a small jet of the desired vaccine into the arm of the patient, the force of it breaking the skin without touching.

After this, I and a few others were sent to a separate station where we were given something different from the crowd. I needn't have wondered when the investigational drug would be tested- this was it.

The onset of the effects of this drug was fairly quick, and was quite distinct from the discomfort of the vaccinations. The intramuscular administration got it into the bloodstream quickly. At this point in time, very few people had ever had this experience; a few mental patients in a hospital in Montreal, in a CIA experiment supervised by Dr. Ewen Cameron, the Agency being constrained from operating in the US, and some people in the Agency itself. There are stories of agents surprising each other with unexpected doses. The US Army was also involved, probably at Edgewood Arsenal in Maryland and in a well documented study in San Francisco around 1959, where an Army Major supervised dosing unsuspecting "Johns", with researchers studying the ensuing proceedings through a two-way mirror.

Of course, these were things I learned about only later in the sixties.

I felt untethered, a little disoriented. Immediately someone came over to me and started asking questions. He told me to come over to where there appeared to be a clothesline- metal tubing with "clothesline" strung between the posts. One effect of this medication apart from a feeling of uncertainty, was a heightened sensory awareness. I could see immediately the line was electrical wire with AF grey insulation, and also intuited that I had been maneuvered to a hidden microphone.

He began to give examples of anti-government sentiments and asked if my father had ever expressed anything like that. I answered no, and he asked what about my uncles, again no, and asked had I ever heard anybody say these kinds of things. Well, my relatives, all veterans of WWII, had not, but I had assumed that anti government sentiment had been expressed by nearly everyone else in the country, particularly by right wing types, I had in fact read some things like this at a conservative bookstore. Given how I was feeling, however, I was unable to engage in a nuanced discussion, so I assented to having heard these things.

Later that day, I found out I had been rejected for Meteorology or ATC, I believed then, for security reasons, although it could also have been because of the investigational program I was in. For what I was likely to experience, there could have been no better preparation than the preventive medicine program, which would be my assignment if I did well in the basic medicine course which would follow my successful completion of basic training.

Back at the barracks, several of us complained of odd symptoms to the sergeant, but we were told it was probably a reaction to the vaccines, or homesickness.

I was given two chances to alter the path I was on, though not necessarily the pharmaceutical one.

My test scores were good enough to be offered a job in administration, which at beginning level was concerned with payroll. This came with an offer to skip the rest of basic training. Just before I was given the offer, a couple of us saw a Special Ops airman walk by, with his trousers bloused into his boots, and a special insignia on his uniform. Though none of us were likely to achieve his status we were duly impressed anyway. When the offer came, I was reluctant to give up the rest of basic, as much as we may have complained about it. We wanted to be real soldiers. Sometimes I wonder what my future would have been, had I taken the offer.

I think some older officers may have remembered my father's years in the Air Corps, I was offered a slot in the next week's music flight. The music flight was for those who would be in the service's bands and music ensembles, which often played for functions in D.C. At Lackland they provided bugle calls and a marching drum and bugle corps. I was tempted, but I would have had an extra week of basic training. The reader can see I was of at least two minds about basic. After leaving the service, I ended up in music, anyway.

Behind the barracks was a large open area, the parade ground, several athletic fields, a track and a static display of two WWII bombers, a B-24 and a B-29. Dotted around were a few "Dairy Queen" type snack bars, with window service and a covered patio with picnic tables.

We had timed mile runs scheduled. After all the marching we had done, most of us managed to meet the benchmark, which was something like eight minutes, twice the recently set record of four minutes.

Too Many Push-ups

~

One of our number, a barber, who carried some extra weight, was unable to meet the standard. The instructor singled him out for a lot of verbal abuse, and made him do a lot of pushups. When he faltered, he was made to do more. This seemed to go on for an eternity, and when he was unable to continue, he got up and couldn't move his arms. He was sent to sick call, and we heard later that he would be discharged and get disability payments since he could no longer work as a barber.

Later that week a Puerto Rican recruit had a breakdown. I guessed it was a drug exposure. Staff assured us he was merely homesick, but he was unconsolable. He eventually left.

We had some free time that Sunday, after Mass. There was a civilian style cafeteria across the four lane divided road that split our base in two. When I entered I noticed a row of sundae glasses filled with different colors of Jello. I was told they were only for display, but, for whatever reason I insisted on a particular one and retired to a table to enjoy this civilian treat after a week of mess hall chow. After I left and was walking around the base, I felt this now not entirely unfamiliar feeling- the same as after the immunizations. I was probably being observed, since this was the point of the drug investigation, but I was not aware of it at the time.

The Rifle

~

During the next week, we were introduced to the rifle. We airmen were not assigned a weapon. As needed, we would be provided with one, that would be taken away after we had finished the activity. First we had to master the "Manual of Arms". This was a series of commands, a drill which governed the handling of a rifle in formation. We were lined up in front of the barracks and issued rifles.

We learned the positions you hold the weapons; right shoulder arms, left shoulder arms, port arms, present arms, and parade rest. We practiced marching with arms, and the transitions between the commands. Later we would learn the firing positions, prone, sitting and standing.

A couple days later, the most intimidating part- disassembling, cleaning and reassembling the rifle.

We were in the barracks sitting on the floor with a towel to put all the parts on.

The weapon involved was the M-1 Carbine, not the heavier M-1 "Garand" our fathers and uncles used in WWII. The Carbine was an officer's weapon in that war and in Korea. There were two versions, one with a wooden stock and a collapsible model with a metal stock for the Paratroopers. Our version had three settings, single shot, and with the appropriate magazines groups of three shots and fully automatic. Ours had a ten shot mag. So on three shot setting you would have three bursts and a single. We never used that setting.

I remember thinking I would never be able to complete this task. We started by removing the bolt, then separating the stock from the metal part, so far, so good. With expert instruction, the rest of the disassembly proceeded logically. We were shown how we would clean the bore after every use, and were warned that neglecting this could cause the thing to explode, endangering ourselves and those around us.

Fresh from disassembly, we reassembled it. We practiced this interminably until we could do it with certainty and ease. We were proud we had accomplished what had seemed impossible only hours before. Then we got the bad news. Having successfully worked on something with 15 or 16 parts, we now had to take apart the bolt. The bolt on a non-automatic rifle might have three or four parts, but the automatic fire assembly was entirely in the bolt. This thing the size of a small cigar had like forty parts.

This took somewhat more time, but luckily it broke down into subassemblies whose constituent parts fit logically together.

Today, the rifle is available only with a single shot bolt, the way it was during WWII.The automatic version is illegal.

Later that week, I was sent to the green monster, alone, for some further testing. For color blindness, as it turns out. As I sat down in the office where I had been sent, I was offered a cup of coffee.

Of all the about weekly doses of what I came to know as LSD, the majority were in coffee.

I was struck by the extraordinary beauty of the test charts, 4" circles containing many small spots of varied colors in which were embedded numbers formed of spots of a contrasting color. These were arranged such that someone with color blindness would not perceive the number. To my eyes these were small pointillist masterpieces. I noticed no ill effects from the medication. I believe that the dosage was on the order of 85 micrograms, as opposed to the typical "hippie" dose of a thousand mcg., the dose marketed by Augustus Owsley Stanley, the supplier in the late sixties to the Grateful Dead among others. As was often the case I was given some time off afterward, which I spent walking.

The next weeks were spent in the field, sometimes marching back to the barracks at night, and once overnighting in the open air. For this occasion, a field kitchen was set up, and we slept in a long row on a plastic tarp.

One day, we marched out to an area near where the field exercises were held, past a Eucalyptus grove to a building with a room the size of a large classroom. I was seeing Euc's for the first time and noticed their baroque irregularities.

We had been supplied with gas masks and were told to put them on. We then filed into the room, maybe 90 of us. A canister was pumping out tear gas, and after a while, the people in charge added another canister, making some fuss about it, adding what turned out to be an aerosolized form of LSD. We were then told to remove our masks, and in a test of our discipline, made to wait for some time in the building. It was something like ten minutes, then we filed out in an orderly manner. The fresh air was a real relief, and we were directed to wash our eyes at a set of taps outside the building. The water carried the LSD into the mucus membranes around our eyes, and into our system; it was a good source of administration. Some for whom this was their first exposure were alarmed, but I knew what was happening and was able to look at the Eucalyptus with new eyes, marveling at their lurid shapes and what people later in the sixties would recognize as psychedelic colors.

Another day we had an obstacle course. The first thing was a low structure of two by fours under which we would crawl a distance of maybe fifty yards as machine guns fired above our heads. There was some skepticism as to whether it was real live fire, but I wasn't going to be the one to test it. Later, a twelve foot high wall was presented to be scaled by us. Quickly we figured out it was only possible if one stood on the other's shoulders, after all had gone over but the last one, me, I was taken aside and offered, guess what?, a cup of coffee. Two or three training personnel stayed behind with me and waited for the drug to take effect. As it did we were at the bottom of some rappelling ropes hanging off a cliff. I misinterpreted that I was supposed to climb them- everyone laughed when it was explained. Instead I was to enter a building full of smoke, go through several rooms, keeping low to avoid the smoke, then exit through a window. Next I was to negotiate some rocky uphill terrain and then rejoin my group.

The Fraternity

One afternoon, after the usual structured activities, my sergeant told me to sign out, and take the afternoon off. This was completely unprecedented- no trainee got this kind of time off. I had gotten some slack after some activities, just some time before having to report back to the barracks. I can't remember if it was suggested to me, but as the daylight was beginning to fade

I went over to the parade grounds. I ordered a drink at one of the snack bars, I was the only one there. I was enjoying the cooler afternoon air and became aware of the onset of the drink's psychoactive contents, and watched as a glorious Texas sunset unfolded in the western sky.

I sat listening to music from the loudspeakers, then mixed with the music were instructions to go to an address on a doorway in a building within sight of where I sat.

The building was a Bachelor Officers' Quarters- a BOQ. I knocked and was admitted. There were five or six guys, all older, who asked me to sit down, and engaged in some small talk.

One said," If you could save the lives of twenty thousand soldiers would you do it?

I tried to object- I hadn't joined the Air Force to be a sniper. One said it was the same as having a really accurate bomb. One said "you knew what was involved when you joined."

The first officer said "It involves killing the head of a country". Another said "the head of a nation". I figured the target was Fidel Castro. I had been pressured into accepting, and said "you would pull us out afterwards?" They said they would try, but I might have to sacrifice my life.

They then put me under hypnosis, with the post-hypnotic suggestion that nobody else would be able to. And other suggestions that I still don't remember.

They then issued some direct orders, among which was that if anyone offered me something to drink, I should take it. They swore me to secrecy, which was probably part of the post-hypnotic suggestion, and told me not to try to contact them, they would contact me.

"Don't call us, we'll call you". The session lasted until dark, but I have little to report of it.

Now I had a secret, and had to face the very real possibility of death, like any soldier.

I walked back in the darkness to my barracks, signed back in and joined the rest in slumber.

Soon after, we were out in the field again. Our platoon had set down their packs for some activity, and I was assigned pack guard duty. Of course, I was not allowed to smoke on duty. Someone, a latino, came by with a pack of Mexican cigarettes and offered me one. Here is an example of conflict in the chain of command. I had been ordered by a low level NCO in my direct chain of command not to smoke, but I was offered something, which according to higher officers, I had a standing order to accept. Naturally I smoked it and got a buzz as it had been doctored with some form of cannabis. We had been taught by our drill instructors not to apologize, nor to offer excuses, so when the platoon returned, I simply had to take the grief I was served by my supervisor.

On the weekend after our first paycheck,(we earned about 12.5 cents an hour, the same pay a carhop earned, before tips. We, however were paid 24 hours a day, 7 days a week, so our pay was about $85 a month),I splurged on a major hotel downtown, next to the Alamo, and an appointment with a tailor. The tailor visit yielded, two weeks later, a summer dress uniform, not part of our regular issue, the same cut as our wool blues, but tan polyester.

As to the hotel, the relief from communal living I sought was not to be. I had already checked in before visiting the tailor, and as I was walking around a young man about my age engaged me. He said he was in the Air Force and suggested we go to a restaurant he knew. The waitress or proprietress behind the counter seemed to know him and he said he was a local. He commented that he had expected me to discriminate against him and I shared that I had had a Mexican-American friend in junior high, whose parents were seasonal workers. When he found out I had a hotel room, he insisted I let him stay overnight on the floor. I mulled the possibilities: one, that he was tasked by AF with keeping an eye on me; two, that he would rob me; or three, that he thought I was a likely pick-up. He had showed me around town with a local's perspective, showing me the Catholic church favored by the Latino population, and the waitress was some relative, so I assented. The evening passed without incident and he left early in the morning.

Not with a Bang

~

It came time for us to go out to the firing range to qualify on the M-1 Carbine. The range was set away from the main housing and exercise portion of the base. We piled into a flatbed truck with side rails for the ride. We were met there by the range staff, lectured, and issued our weapons. The range had about twenty "lanes" with separating barriers at the pads where we shot from. The distance to the targets, was 50 yards, except for two at the far left, which were 200 yards. We were all lined up at the 50 yard targets when two men came up and brought me over to the longer targets. I was offered coffee again. One of the staff came with a different rifle with a telescopic sight, tested it and adjusted the 'scope. By this time I was feeling the effects, and understood this was research as well as my certification.

We would score one of three levels, marksman, sharpshooter or the highest, expert.

I was to make all my shots from the prone position. By the time our targets of record were up, I was in a heightened state of attention and could feel my heart beating. Before firing, I was told to squeeze the trigger in between heartbeats. Looking through the 'scope, I saw the crosshairs shake on each beat. Lub-dub, lub-dub, the triplet rhythm of the heart, one two, one two; after the two a space of calm. In this space squeeze, not pull, the trigger. Watch where the shot goes, make the minute adjustment, watch the next one. Hard enough under controlled conditions, imagine it with the stress of battle, or the high stakes of sniper fire.

I was getting into a rhythm, all the noise falling away. hearing only my heartbeat, and, faintly, my coach's suggestions. I was three shots away from an expert rating- 300 points. My coach says "miss the next three shots: a heartbeat's hesitation, then I follow the order. I will earn the sharpshooter medal, a silver Maltese cross with two concentric circles in the middle with a little bar hanging under with the word "Rifle". The only medal I would be awarded.

The truck ride home: still a little high, and with a great sense of accomplishment. The afternoon breeze drying the sweat of the day.

Reflecting on the eight weeks of basic, I see now that I went from being an individual to being a part of a team, as much as that is a cliche. It's more like we were a single organism. Waking and sleeping at the same time, having our meals together, moving about the base like a two hundred legged caterpillar. We were soon to be a hundred blue winged butterflies, moving into different groupings for further training.

Things that stand out in my memory are; the first time I wore my summer dress uniform- I had picked it up on a Saturday, and the next day, I put it on for a Sunday stroll. With no stripe on the sleeve, it could be mistaken for an officer's, missing only a little gold bar to identify its wearer as a second lieutenant. A new recruit made that mistake and saluted me. Not knowing what to do I returned his salute. An officer nearby came and chewed me out, saying that I could be prosecuted for impersonating an officer.

The afternoon off, with its remarkable sunset, before I was initiated into the fraternity I could not acknowledge, the spiked jello, and my time at the firing range also stand out.

An odd memory, of being administered a Rorschach Test. We came to one inkblot that resembled a butterfly, though on further examination was a nude woman sitting with her legs apart, knees up, a typical porn pose. For whatever reason, I would not acknowledge that's what I saw. I've successfully repressed the memory of KP.

When we began our rank was E-0, or Recruit. Now we were airman basic, E-1, with a corresponding raise in pay- still no stripe, that would come after our second phase of training, when we would become airman third class, E-2 and would sew the one stripe insignia onto all our uniforms. For our fatigues there would be a pin for our caps to mark our status.

Our graduation, and a parade past the base commander was the order of the day for our's and several other Flights.

We were up early to prepare and all looked sharp. We marched to a hangar on the flightline where the pass-in-review would take place, and waited in the cool shade inside for our marching orders. But, one more surprise was in the cards.

The heat of the runways on that late August day required hydration- someone fainting in formation would reflect badly on our NCO's, so we were issued canteens and told to fill them. Some guys looked inside and there was a powdery substance visible. We were assured nothing was amiss, amid some grumbling. I, and a few others knew what was up, and was prepared to make that extra effort to compensate. I was feeling positive and anticipating a great day. However, nearly everyone in the flight would be in this experiment.

I started feeling the effects, then suddenly a large number, perhaps having their first LSD experience, started panicking. The "bad trip" which became familiar during the "Summer of Love" was in full swing. The NCOs who were in on the game tried to calm down those who were having a hard time.

I was still raring to go, but our participation was cancelled, probably excused by an outbreak of "illness."

So Basic Training Flight 866, Summer of 1961, Lackland Air Force Base, San Antonio, Texas, came to an end, minus only two recruits, and not with a bang, but with a whimper.

Medics

~

The East-West highway cut Lackland in two parts. South, the barracks and fields of the basic training flights. To the North, Administration and dorms and classrooms for advanced training.

From 866's barracks, turn right on the highway going East to a large boulevard, another right turn takes you Downtown. However, a base bus ride to the Southeast Gate brings you to the Mexican-American side of town. From there, a large wooded boulevard, an "Alameda", led downtown. Anglo servicemen were reluctant to use this shorter route, but my introduction to this side of town made me confident in using it. Respect engenders respect.

This day, though, I was crossing north with every possession in my duffel, and reporting to my new dorm. It was on a quadrangle with dorms on the north and west sides, classrooms on the east, and on the south, temporary housing for those processing out. The first instruction we received was to not harass these folks, not to be too proud of ourselves for making the cut.

It was like a college dormitory; single beds, not bunks. Smaller bathrooms, cleaned each time we used them, not cleaned by a work detail as in the barracks. Each bed had a nightstand and a lamp, and we had dressers, not footlockers. We wore our khakis, and were issued fatigues only for the rare work details. We wore our Oxfords and stowed our Brogans. All in all, a marked difference from basic.

Daily coursework was from 8:30 to 4:30, without a lot of homework. The certificate awarded at the end was the equivalent of Licensed Vocational Nurse.

The first things we were taught were the limits of our authorized activities and how we had to cede most responsibilities to the M.D.s and R.N.s. We did what they told us to.

We did not have to wait for permission to practice first aid. The important things were to stop the bleeding, maintain open airway, and treat for shock. The first two were of equal importance. Hand pressure and tourniquets for the first, manually clearing the tongue from the throat if it had been swallowed or a tracheotomy, if necessary, for the second. Treating for shock meant mostly keeping the patient warm. We were especially cautioned not to move the victim if there was any chance of spinal fracture. Later, we would have special sessions for combat medicine.

We learned basic medical terminology, prescription procedures, and basic hospital procedures. We practiced taking blood pressure readings, heart rate, and body temperature, and entering these in the charts.

We learned the importance of procedures to prevent and relieve bedsores in long term patients.

In our class, we had one woman student. Officers had to caution us to respect her as much as any other member of our class, but nothing untoward ever happened, as an atmosphere of camaraderie prevailed.

Time came for field exercises in combat medicine. Some Airmen took the part of the casualties and had rubber prosthetic wounds attached to their bodies. Others of us were issued packs with supplies. Our classwork had covered various wounds and their field treatments, splints for bone breaks, and emergency one dose opiates- a little tube like a tiny toothpaste tube with a short needle which the medic would inject into a muscle for pain. These were, of course tightly controlled. For the purpose of our exercise, we would just announce we had injected our patient.

The cruelest of our lessons concerned "Triage": sorting people into three main categories; those who needed minimal care and could return quickly to the battle, those who needed evacuation to the rear, but had a good prognosis for recovery, and those who had such serious wounds that they likely would not survive.

In military doctrine, the first category got the first attention, followed closely by the second, though the second might receive care first depending in the wound. The most seriously wounded might get only palliative care. We had tags to attach to the wounded.

I had received a dose of LSD prior to the exercise. Even though we knew the wounds weren't real, the atmosphere felt more realistic than we expected, and emotions were high as we heard the "victims" crying out in pain and moaning. The wound simulations were unexpectedly realistic and I can say we all took this more seriously than any other training we had been through. We all did our best carrying out the steps we had done in class and were in a somber mood afterward.

Once we had a work detail where we had to move furniture from one dorm to another. We, or at least I had been dosed and we proceeded to carry items along the paths. As I stopped momentarily to look out the third story window of the first dorm, The airmen moved with a coordinated rhythm and again, I saw us as a single organism.

After we passed our exams, we were awarded our certificates and promoted to airman third class, the third ranking in the enlisted Air Force. Like a third class rating in the Navy and the Private First Class in the Army, our pay grade was now E-2.

On one weekend, someone, probably from my unacknowledged fraternity, took it upon himself to take me over to to the Kelley side of the base. Kelley, unlike Lackland, had runways and he wanted to show me some flying. The folks in charge were reluctant to give him a plane so, instead, he took me over to a confidential training area. A southeast Asian village had been reproduced, down to a beautiful black-haired woman in native clothing outside a house on the edge of a stream- an idyllic rural scene. A path led past the house and over the stream. I was told to walk the path but not interact with the woman.

Only later did I realize that I had seen preparations for our further entanglement in Vietnam.

We received our orders for our next assignments; mine was to Preventive Medicine

School at Gunter Air Force Base, now Gunter Annex, Maxwell AFB, Montgomery, Alabama.

My orders gave me four days and a weekend to get to Montgomery, and I booked a small compartment on the train to St.Louis, ready to enjoy a moment of privacy at last. There would be none at home with my parents, my brother and three sisters.

Most high school friends were gone, in the service or at college. Charlie, who had joined the Catholic Church with me, had joined the Army a year before my enlistment. He took basic at Ft. Leonard Wood, some hundred miles southwest of St.Louis on Highway 66. The base was named for a hero of the Spanish American War in the Philippines. He would be seen today, by some, differently.

Charlie had finished basic and stayed on as "cadre," training the next crop of recruits. After that he went to Ft. Devens, Massachusetts, and then on to Korea, duty he found congenial, and came back with a love for Korean language and culture.

I took my duffel and a guitar and rode a Greyhound bus to Nashville, transferred there to a bus to Montgomery, via Chattanooga and Birmingham. The experience of traveling through the south was a little unsettling- I was warned by my my mom at about 11 years old not to go to the south, where our ancestral lands were wrested from us in 1837. I was insulated by being in the bus- Interstate travel by bus and train had recently been desegregated and there was no trouble. Soon I reported to the dayroom at Gunter and was driven to my new billet.

In addition to my Preventive Medicine class, the dorm housed some people who had skipped basic.There was an older guy, a civil engineer, who got his own room. He was doing his six months of active duty before being transferred to the Individual Ready Reserve, where he would serve the balance of ten years. During this time he could be called up for temporary projects at any time. This served the AF's need for professional services.

Three or four other guys were sons of diplomats or other important people who had been allowed to serve without going through basic. I never found out what their nominal jobs were to be. Maybe they were part of the fraternity.

I began a very intensive sixteen week course. Class was eight hours each day with homework at night. Venereal disease was one of the first subjects; discussions of the main three, with training films showing each's ravages, along with the principles of prevention and how to do interviews for contact tracing. AIDS was unknown at this time. We had units on animal diseases that could affect humans, principally tuberculosis, which could be contracted from dairy products from infected cows and anthrax, primarily carried by sheep. Mention was made of the biological warfare potential of anthrax though it would be covered later under the general rubric of CBN, chemical biological and nuclear warfare.

We learned the concept of the vector- an animal or insect that could be an intermediary between other animals and humans, or between humans, in the spread of disease. We studied measures to rat-proof buildings and screens and nets against insects. We were made aware of the tradeoffs involved in the use of pesticides-you killed the bugs but also exposed people to neurotoxins.

We studied measures to limit exposure in agriculture, and also looked at exposure to toxins in industry, particularly in aviation. We learned to collect samples of air and water to send for lab testing for these toxins and infectious agents.

One of our longest projects was to become proficient at identifying insects. If breaking down a rifle was complicated, being able to identify every member of the orders of insects seemed overwhelming. Our main tool were the "keys," books that broke down the distinctions; kinds of wings, kinds of mandibles or snouts, etc. By following prompts like 'if wings have covers go to page 235", we could soon begin to ID various families and with details come up with the exact species. We would go out in the field and collect as many as we could find and come back to the lab, euthanize them with a drop of ether then mount them to a board with tiny needles, "minuten nadeln." Then came the work of identifying them. Finally the seemingly interminable work with insects came to an end.

Chemical, Biological and Nuclear

~

An even more serious chapter of our training was next. For the work with nuclear information and materials, we needed new ID's reflecting our newly acquired Top Secret, Nuclear, Compartmentalized clearances. The compartmentalized part meant that we were allowed access only to nuclear information, not the rest of the top secret category. Non- compartmentalized top secret clearance is fairly rare.

A little badge with photographic film inside partially shielded by lead was issued to workers anytime they are around a source of radiation, including ourselves. These were sent each day to a lab for development, just like black and white film. The contrast between the shielded and unshielded section was reported as a number in "Roentgens" that measured exposure. We learn to record and report that number, enabling supervisors to monitor workers radiation dose.

We had lectures from doctors summarizing the dreadful facts gleaned from the Hiroshima and Nagasaki blasts. The next day a general came and gave us the official position, which was that exposure was not a big deal and that the US might have to use these weapons again. It wouldn't be too much exaggeration to say his attitude was "Just walk it off."

We learned to calibrate and use Geiger and alpha counters. These were in the popular imagination because prospectors would use them to look for Uranium. We turned on the Geiger counters and could see the "background radiation," a level caused by materials in the earth and by cosmic rays from space. These levels would be disregarded in measuring higher levels. To calibrate further you needed a sample of a known strength.

For this purpose, a small piece of plutonium, encased in plastic, was stored in a lead lined container under the building. This was brought up and passed around to each of us. We were told to keep it at arm's length. To test the two versions of the danger of nuclear material, I slowly brought the sample closer to my head. As I did, I could actually feel the radiation as it got closer. My supervisor yelled at me to put it farther away, which I did. I decided to believe the doctors over the general.

We had a field trip to a dam and water treatment plant northeast of town that supplied Montgomery with drinking water. It was an uncomfortable ride in school- type buses painted AF blue. After a tour of the facilities including the final station where technicians would take samples for lab analysis, I was lingering outside the door to this room when I heard officers saying that if you were to add LSD to the water supply, this is where you would put it. My ears perked up and I walked back into the room. The officers quickly changed the subject, and we were offered lunch outside on the lawn. I ingested the drug from something in my lunch, but didn't feel it immediately. We got back on the buses, As the effects begin to unfold, I have the most pleasant ride back to base, glimpsing picturesque little towns and beautiful countryside from my window seat.

There was a State Prison across the Highway from the main gate, and a dairy that served it. We were to learn about sanitation procedures at the dairy, and we all trooped over there in our clean uniforms and shiny shoes. The barnyard, however, was three inches deep in mud the consistency of pudding. Whatever the lesson was, it was overwhelmed by the task we faced afterward of cleaning our shoes and changing our uniforms.

One day we were told that there would be a short electricity blackout on base. It turned out there was to be an execution at the prison that day, and that they needed all the power they could get. We were told not to joke about it, and this lesson about the state's power over life and death was taken seriously.

About this time, I saw how lacking in rigor the LSD testing program might sometimes be. A small squad of us were taken to an area near where we had gathered insect specimens. The two researchers set off a few smoke bombs of different colors; we were told to run through the fog and encouraged to breathe it in, especially the red stuff. We did so, without much apparent effect. I was learning the the effects were often delayed, but apparently the staff hadn't got the memo. In their effort to get the test over with, they told us to "act crazy" and we complied, doing silly dances and gesticulating. Now, we had learned in class the proper way to conduct medical trials, and having a predetermined outcome was certainly not included. I sometimes wonder if the whole effort ever yielded anything of scientific value.

They gave us some free time afterwards, and I had a pleasant walk around the base, enjoying what mother nature had on offer.

To get from the base to downtown Montgomery, I turned left at the main gate, crossed the highway and soon sidewalks appeared. I chose this side of the street since on one of my previous walks I was told, as I passed the entrance to the Stadium where the "Crimson Tide" had their home games, that it would be dangerous to me to pass by on game days. The Chickasaw Nation occupied, before removal, a bit of the northwest corner of the state. The tribe had been split by state borders: Tennessee, Alabama, and Mississippi. I was mostly tolerated, even in the Museum of the Confederacy, for two reasons; one was that there was a largely denied genetic component of Indian and African blood among the so-called whites, and two, it was apparent I was an Airman, Montgomery being home to Gunter, the training base, and the "generals' school," Maxwell AFB. There was a park downtown near Maxwell, and catty corner, the fanciest hotel in town. On the street across from this hotel an airman told me that if I were to enter it, I would be unlikely to emerge from it alive. I did not feel tempted.

One of the first things that struck me on the sidewalks of town were the brown splotches where the results of the use of chewing tobacco reflected the upward or downward slope of the pavement.

I passed through the segregated parts of town, where I could hear blues music coming from establishment where I would not be welcome: not white enough for the whites, nor black enough for the blacks.

The downtown business district reminded me of Atlanta as portrayed in "Gone with the Wind." I wonder if it had been used in the movie.

There was a music store, like its counterparts in St. Louis, where I was welcomed. I bought two guitars, a Gibson dreadnought and a Framus classical from Germany. From now on, in addition to my duffel, two guitar cases would follow me home, and to France.

Toward the end of the class, we had an IQ test. I had a test in junior high, where I scored 135. I suppose I had a reputation, undeserved, as a juvenile delinquent,(I smoked between classes in the bathroom), and the counselor called me in and said "I'm not sure how you did it but I'm sure you cheated." I was flabbergasted. My whole family had always valued knowledge and I spent a lot of time in libraries, both school and public. I strongly denied any shenanigans, and my score withstood the scrutiny.

Our instructor announced we would be part of a study on the efficacy of coffee. Some of us would get one cup before the test, and others two. I could guess which group I would be in.

This is something about the military, since this test would help determine your future, it should be fair and unbiased, but here, the needs of the individual were clearly subservient to the service's agenda.

We got a bathroom break, then we were served our coffee. My second cup had an extra ingredient. The questions were mostly multiple choice.

As we began, the effects of the second cup became apparent. The familiar feeling, but a bit more intense. The walls seemed to be breathing, but I didn't let it distract me. Actually the coffee helped me concentrate and I got into a good rhythm with the test. Multiple choice was an easy thing for me and my teachers at Normandy High had taught me the trick of eliminating the obvious misfits.

Bits of text and questions about their meaning. Geographic knowledge, grammatical multiple choice, word meanings, and math. My high school algebra was OK, but I had done well in geometry. The most difficult part, to my mind was a series of more and more complex geometric figures. One view was shown, then you picked the figure that matched it from a different angle. The difficulty was progressive, and each success prepared you for the next level.

Putting one foot in front of the other, I got to the end of the test. Soon we would find out our scores, and whether one or two cups was better. If my score was good, I couldn't claim credit since my performance was chemically enhanced. My score would have an asterisk after it.

A few days later, our scores were posted with a code which preserved everyone's anonymity. Mine came out 139, four points higher than in junior high.

I guess I learned something in those four years, especially how to take tests.

Meanwhile we had our final exams. I scored number four in my class, and on this basis I got fourth pick of overseas assignments. The list was posted in the library. There was one slot in Great Britain, two in Germany and one at Evreux-Fauville AB in France, followed by assignments in Japan, A bunch in The Philippines and some others elsewhere in Asia. In retrospect, Japan would have been a good choice, but I had taken French in HS, and my library research yielded the fact that Evreux was in Normandy, the namesake of my high school. I knew a choice of England was beyond my reach, and I knew that Germany was considered a plum, so I chose France and was assigned there.

I think they let us go a little early, just before Christmas.

I put my duffel and two guitars in the baggage compartment of a Greyhound and checked them through to St.Louis. There would be a change of buses in Jackson Miss., where I arrived in the afternoon.

This was the time of the freedom riders, though being insulated somewhat from events I was only vaguely aware of it, and of the Montgomery bus boycott- I hadn't watched any TV the whole time I was in training.

When I got off the bus in Jackson, I had a little time before the St.Louis bus left. The station was at some remove from the downtown, which I could see five blocks away. I was wary of going in that direction and meeting some klan types; I wasn't in uniform, which would have given me some confidence.

Though Interstate travel was no longer segregated, the Greyhound station still was. The main building for whites, and a shack in back for the "colored." I faced a Hobson's choice; which one? I opt for the colored office to be safe, but the people staffing it are upset to have me in there, so I intuit that by local standards I'm closer to white. This is in the back of my mind, I'm used to passing for white, to the point where I emphasised my father's Baltic heritage over my mother's native and Welsh origins, so I get my ticket punched in the main office.

As I'm waiting in my seat, a young African American man comes over to the bus by my window and gets my attention but quickly, the local constabulary hustled him away. I hope he didn't face serious consequences.

The driver knew I was a G.I. from my checked luggage and during the trip invited me to sit up front to talk. He explained some of his safety measures-stopping completely at railroad crossings, opening the door and listening carefully before proceeding. He spoke with resignation about the racial situation in the South. Of course it was and is not only about the South.

La Belle France- the Good and the Bad

~

After a few days at home, I got on a Continental Trailways bus to McGuire AFB in New Jersey. Trailways marketed itself as a premium service. Most Greyhounds made frequent enough stops not to need onboard bathrooms, Trailways made fewer stops and had airline-style facilities, a small lounge area in back and had a Stewardess come aboard at our first stop, Indianapolis, who served coffee and sandwiches. We stopped in Pittsburgh, and traversed scenic woodlands across Pennsylvania.

An Air Force C-54, a four engine propellor plane, took us across the Atlantic, stopping in the Azores for refuelling, touching down at Le Bourget Airport in the morning. An hour's drive brought me to my assigned dorm at the hillside compound of the 317th Tactical Hospital, above the town of Evreux.

Coming up to my building, some guys were hanging out on the main steps. One was saying he's just back from Vietnam. This was the first time I ever heard of it, it is hard to realize that at that time it was not a household word with all the connotations it carried later in the sixties. He said, whatever happens, do not go there. He says there is no obvious battlefront. You could be anywhere-in town or in the countryside and you can be ambushed, or a bomb might go off in a cafe as you are passing by. An officer hears this, comes over and tells him to shut up.

Next morning I went to my duty office a few miles away on the main operations base. I met my supervisor, a major who headed the base veterinarian's office out of which we would do our preventive medicine technicians' work. He had been informed that I spoke French, and immediately commandeered me to go talk to his mechanic in town. The exercise is futile; in addition to barely being able to understand him, I don't have the automotive vocabulary to even attempt to convey the major's questions. Off on the wrong foot, already, but the major was not too miffed.

The base is a transport hub. C-119s, flying boxcars. Our newest planes are the C-130s which after many iterations and improvements still fly today. In 1960 the base participated in evacuating civilians from the Belgian Congo and the airmen who had been on base at that time wore campaign ribbons commemorating the operation. While I was there the main aviation activities were touch and go landings for the officers to maintain their ratings.

The base began as a grass airfield for an aircraft factory in the twenties. During the thirties it became a base for the French "Armee de l'Air". The German Luftwaffe built concrete runways and secret tunnels underground during WWII, after taking over the town that they had bombed extensively. During my time there, many parking lots stood on the sites of destroyed buildings, though a 15th century tower and the cathedral were still standing.

The Nazi base was bombed many times by the Allies. The Germans had built a decoy airfield some miles away, but by war's end it still remained pristine. Now Base 105 of the French Air Force, it houses French air transport activities and services the French president's equivalent of Air Force One. There is also a museum exhibiting uniforms and artifacts of the base's previous tenants.

The daily routine was a trip to the motor pool to be assigned one of several '47 Dodge field ambulances, a 6 cylinder flathead four wheel drive vehicle, painted AF blue with a red cross on the side. Inside, room for four stretchers on benches, two to a side. With the top ones folded up, room for four passengers on the bottom benches. We took turns driving the 3 or 4 miles to and from the base.

My immediate supervisor was a sergeant who had been with our unit when it was stationed in England. He explained the workings of the base vet's office, primarily meat inspection, secondarily to attend to the needs of staff's pets. If we'd had horses their care would have fallen to the major. As a vestige of this duty our office received, every three months, a flat round red tin filled with cigarettes of hemp grown on the government's farm in Mississippi. It was kept in a closet. When the new one came, the old one was sent back. Apparently marijuana was a good horse tranquilizer. As the new one came in, the cigs would gradually be replaced by "gauloises," which looked pretty similar, by intrepid staff. When I was shown the can, there was no pot left.

I was due for a little hazing. Some NCOs sent me to an officer's house near the hospital compound to take water samples. Only the officer's wife was home. Apparently she had a rep for sleeping around, and I was put up to say something to her, I forget what. Being the follower I was I said it provoking a reprimand later, though the NCOs that put me up to it got a little worse. Anyway, the newcomer was initiated.

The duties of our office were mainly food service inspections. We would check out kitchens at mess halls and canteens ensuring that food was kept at appropriate temperatures "Cold foods cold, hot foods hot," and to ensure there were no pest infestations. We monitored "sell by" dates on "C" rations. A perk was that each ration came with a pack of cigarettes; Lucky Strike or Camels, unfiltered, that we kept. We also examined and sampled the boxed flight lunches for the air crews. We were the King's tasters. Every inspection would be documented by a short report. The report was seen as the most important part, in the same way that the count of the cigarettes in the can was more important than their nature.

We also inspected barber shops, but we didn't bother the Cantine Francaise with its unrefrigerated cheeses, the favorite eating spot of the senior officers.

Finally, we monitored reports of VD infections and the locations implicated. We did our part in the chain of contact tracing, interviewing patients and reporting their contacts.

The town of Evreux was fairly small, a walk from the city hall to the hospital (the Fauville part of Evreux-Fauville Air Base) took maybe forty minutes. The walk to the operations base was considerably longer. Behind the buildings on the main street was a small river, the Iton, that ran through town. From the cathedral to the hospital this was the scenic route, which passed the city hall, a smaller church, and a 15th century clock tower. From the edge of town, near the bus station, a road wound up the hillside to the hospital compound. Entering, to the right were the officers' quarters, across from them the hospital building, then the canteen with its slot machines and a beer bar. The motor pool was next, then the two dorms for enlisted personnel. At some remove were suburban style houses for higher ranking officers.

My work week was 9 to 5, Monday through Friday, and when I had the chance, I went to Paris on weekends. It was 60 miles east; an hour by train, a bit more by bus and a couple hours hitchhiking.

This was before the route had freeways, and every intersection had low concrete markers pointing out directions. Simply follow the signs. You could get easy rides to the edge of town, walk a few minutes to the other side, find the sign and get a ride to the next town, and often to the outskirts of Paris where you could take the Metro.

There were a couple of routes, one through the town of Mantes-la-Jolie, at that time a distinct urbanization, parallelling the train route. Mantes is now a suburb of Paris, with a possible side trip to Versailles, and another through Dreux and Rambouillet, the former Royal hunting grounds with wide avenues cut through where the aristocrats could see game crossing from one thicket to another.

NATO headquarters was at Rambouillet. One time I got a ride from an officer there. He tried to get me into the PX there, but it was only for officers, so I waited for him in the parking lot- it was interesting, I saw NATO officers in their different countries' uniforms, and lots of BMWs,

Mercedes, Alfas and Jaguars. He came out and we continued to Paris.

Another time I got a ride from an older officer who was in on the LSD experiments. He pointed out a Chateau that had been used by the S.S. where interrogations (torture) had been carried out, and was now controlled by the Americans. He gave me some dosed coffee, and let me off at a suburban Metro station, which seemed to exemplify the "City of Light" and encouraged me to take the opportunity to experience the privilege of being so close to a city many only dream about. Under the benign dose I received, I got to know the subway and could get around easily.

On one weekend, I went to the bus station, close to the hospital compound, I bought a coffee from a cart and boarded the bus, a smaller Euro style in contrast to the big greyhounds. I later recognized the the coffee vendor and the driver were in on the "sting," if we could call it that. Again I could tell I had been dosed and after a little initial disorientation, I settled back to enjoy the ride, which ended up on a new stretch of expressway near the city. I was the sole passenger.

Once I determined to go to Brittany, where some of my ancestors were from. Little concrete direction markers, one inscribed "Bretagne," and one, "Paris," stood at an intersection I often passed. As I started soliciting rides a feeling of dread came over me. I guess my plans were conflicting with some kind of suggestion that I was expected to go to Paris, as usual. I went over to the Paris route.

On one of my first trips to the metropolis I went to the Place St. Michel. On the west side of the place, a woman approached me and invited me into a cafe where we had choucroute garni. She was , in retrospect, obviously my contact there and nearly every time I came into town I met up with her. She had a room at an unnamed hotel at #5 Rue Git-le-Coeur, "Heart's Rest Street." She said her name was Penny.

She said her husband or boyfriend was a writer who was on assignment in Algeria.

She introduced me, over time to other denizens of the hotel. It had maybe a dozen rooms, on three or four floors, with bathrooms and W.C.s on each floor. Nobody was ever at the front desk, and when one left, one hung up the key on a board behind the desk. Most of the other visitors were pilots in transit, Americans, British, and Scandinavian.

My sergeant, having been stationed in England would sometimes jokingly use a British accent, as did some of the American pilots. Thinking it cool, I picked up the habit as well.

I still considered myself Catholic, at least with the moral indoctrination of it, to the extent that I didn't put the moves on Penny, given that she seemed to be in what we call today a committed relationship. My new friends, however seemed to encourage me to deemphasize my roots to fit in better with the anticlericalism of the French left, given that the church in France was associated with the right wing. They tried to fill me in on the usual attitudes of the younger generation intellectuals, and to be able to discourse to a degree with atheists and Marxists.

Algeria was a current preoccupation among the French. I think by this time most people were used to the idea of independence in North Africa, but cafe bombings still happened, and there was a right wing backlash in the form of the OAS.

When I first saw graffiti saying " a Bas l'OAS," I couldn't figure out why anybody here was upset with the Organization of American States. Then I was told it was l'Organisation de l'Armee Secrete, the Secret Army org., a group of French officers unhappy With events concerning Algeria.

In this same neighborhood was another hotel with a similar clientele on Rue de l'Hirondelle- the "Street of the Swallows," where I stayed later in '62 and again in '64. Back on Git-le-Coeur one door away from #5 was the "Beat Hotel" where Allen Ginsberg, William Burroughs and Peter Orlovsky, among others stayed. Were they, wittingly or not, involved in the cold war propaganda effort against the Soviets? The U.S. trying to showcase our freer society?

Meeting Dr. Gottlieb

Back to the day to day at Evreux. I was assigned to inspect a bakery. When I got there, I became suspicious- It was just a storage building, one of several in a row, with flour smeared on the steps. I went in and there were two or three guys there, no baking equipment, and a table with a cake on it. A pile of cardboard boxes obscured the back of the room.

They sat me down, and I understood this was no ordinary inspection. They showed me a little bottle, smaller than a half-pint booze bottle, like a large aspirin container or an art supply painting medium flask. About three by three by one inch, filled with a dark blue liquid. The label said Sandoz Pharmaceuticals, "Delysid" in large letters, smaller print saying "solution", and giving a percent concentration.

They told me this is what I'd been getting, and offered me a piece of the cake. It had blue and white frosting, the same color blue as the stuff in the bottle, and they encouraged me to eat some, especially the blue icing. Alice Through the Looking Glass, anyone? I complied.

An elder, red-faced, Eastern European looking gent, who I now believe to be Dr. Sidney Gottlieb, came in was greeted with deference and set up a monitor. The guys bring out a polygraph with its usual hookups and tell me they will be asking some questions. I'm hooked up, a tube around my chest, a device on my fingertip, among other things.They turn the apparatus on and ask a couple baseline questions. I'm a bit apprehensive as they begin to inject Sodium Pentothal into a vein, and I feel like I'm dying. I say "Are you trying to kill me?" They take out the needle. I am asked a variety of things, and finally, "Have you ever had sex with a man?' I deny it at first, but concede, " Yes, when I was a kid." This seems to be the answer they were looking for. They were not interested in hearing the circumstances. An 11 year old neighbor boy said he would show me what adults do. Naturally, having heard of adultery I was curious. At this point, I got the wild notion these people were not authorized and I began answering all the questions with "name, rank, and serial number", looking at the doctor and saying "You're all foreigners!" At that point some American-looking Psych techs and a lieutenant colonel with a .45 in his holster appeared from behind the boxes to reassure me I was indeed under the care of the AF.

I was told never to reveal not only this incident, but anything connected with my service. I felt this was not in line with due process and said I wanted to talk to the base commander. If I did, the colonel said, they would just drop me on the flightline, the rule on the flightline being that unauthorised intruders would be shot. Nothing like a death threat to ruin your day.

There was some discussion as to when I could talk about it, the basic period of confidentiality in the military being twenty years. Thirty years was bruited and the colonel said more like forty.

Someone said it would be better to assume never. My judgment today is that I'll go with the colonel's assessment, and add ten years for good measure.

The secrecy requirement was reiterated and I was given a shot of Thorazine which put an end to the feelings, good and bad, of that day. After I came to and was released, I tried to walk it off, but none of the pleasure I usually derived from walking was available. The worst of it was I felt I had to genuinely make an effort to forget everything, not reflect on it, and basically begin my life as if none of it had happened. I felt depressed for a long time.

I had one more ordeal to go through. Incidentally, there was an Army caserne, as a small installation is sometimes called, at St. Andre about 15 miles from Evreux. During the eighties I heard about a black GI suing the Army for what seemed to amount to torture during a session with LSD they conducted on him there. This was explored on national TV news. He felt intense fear and paranoia, quite different from my experience up to the point of the bakery incident.

During the fifties and sixties the Army, at Edgewood Arsenal or Ft. Detrick in Maryland exposed volunteer subjects to BZ (3-quinuclidinyl benzilate) (sounds bad and it is), seemingly a non lethal nerve gas derivative.

Given the extreme reaction our St .Andre soldier had, I guess he was exposed to this BZ also. That combined with the fraught racial situation of white officers experimenting on a black GI could have provoked the horrendous ordeal.

Twenty years after the exposure, a lawsuit was allowed to proceed against the Army, and the plaintiff prevailed, being awarded a settlement of $500,000.

Unfortunately, this did not seem to bring him peace, and years later the news reported his suicide.

As to my experience with this substance let me reiterate; I volunteered for the LSD study, and would not trade my experiences for anything, but never consented to try BZ. The reader will remember however, that I was under direct order to accept any drink.

After the bakery incident, I was slowly feeling better and my spirits were lifted when I received a three day pass, Friday, Saturday, and Sunday, to go into Paris; maybe I would see Penny and recover some of my momentum.

I went down to the bus station I had used before. I was awaiting the bus and an American came out of a bar nearby with a highball, and offered it to me. I was wary since the bakery and declined but he kept insisting and remembering my orders, I gave in. (I had been ordered to forget all my service experiences, and now here I was under obligation to remember again. The most diligent soldier cannot follow conflicting orders.) I drank it and immediately was sorry I did. I could ordinarily handle this amount of liquor, but something unfamiliar was in it. I started thinking this was not the official experiment I was part of. I had difficulty controlling my muscles- something that never happened with LSD. I immediately gave up my plans to go to Paris, and ignored any further interaction with the guy I believe was not Air Force, but Army or CIA. The CIA was constrained in its operations in the states, but here in France had free reign.

As it was happening I could think of nothing but getting back to my dorm. I struggled up a back path, the old steps of a chateau, which led to the back of my dorm. I knew I couldn't go to the hospital since I might reveal something in my condition so I got in my bed and pulled the covers over my head since I was extra sensitive to the light. I was awaiting to see if I would die, but when that didn't happen, I began to do an inventory. "Stop the bleeding, maintain open airway, treat for shock". I was glad to be able to use my basic med training. I was a qualified medic in any case. No bleeding, still breathing, patient is warm under the blanket, I couldn't feel any internal malfunction so I decide to wait it out. Next morning, I was alone, roommates also on leave. I get up , look in the mirror. Alcohol's effects gone but the other thing is still raging. I looked horrible, everything I looked at looked horrible.I get back under the blanket and only get up to go to the bathroom and drink water.

Sunday evening my roommates got back, I was still riding it out. My mates are worried. Some other people come in (they are probably in on the game) and ask if I'm OK. The symptoms were lessening so I said I had things under control, and they left.

A third night of sleep and meditation brought to an end three days of Hell, mastered only by a self-directed sensory deprivation. In the morning I am able to make myself presentable for work and took the ride into the main base. I still didn't feel well. The world still looked ugly.

One thing I was certain of. With this fractured chain of command I couldn't trust that this wouldn't happen again. I refused go through it again. I only trusted myself and I planned to work out how to get discharged.

The week following my aborted three day pass, was difficult, I still felt weak and barely able to do my work, but somebody in my office must have known what I was going through because everyone went easy on me. I was irritated by my roommates. In retrospect, I was suffering post traumatic stress, something we didn't recognize at that time. I resolved I wouldn't go through that again. My previous feeling that I was treated fairly was undermined.

On one of my long walks in Paris, south on Boulevard St.Michel, I had noticed a building housing the Swedish Consulate. I didn't mind being in the military, just the possibility of being dosed with BZ again. I went to the consulate and spoke to the military attache' about the possibility of getting transferred to the Swedish Air Force. I had no idea if it was feasible. He called someone and had a discussion, but counseled me that it would be impossible, and that I should just try to bear with my difficulties. Of course, I couldn't say what the problem really was.

I wrote a letter to the Base Commander, which while not disclosing anything classified, I expressed my desire to be relieved of duty. I was checked into the base hospital, then transferred to the hospital at our sister base in Orleans. For about ten days, I saw a psychiatrist, but I never disclosed the truth to him.

One afternoon I was in the bed, and the nurse came by with a pill she wanted me to take. I resisted, she insisted. The psychiatrist must have known, but I wasn't going to be the one to tell him. It was suggested that I go to the auto show that was being held in a hangar nearby. Since I was on my way out,I thought it was pointless for me to be looking at a prospective purchase, but enjoyed looking at the new models anyway. I knew I had been dosed with LSD, not the other drug that had caused me such distress.

On the way back, it was sunset, and just as I passed the flagpole where both the French and American flags were flying, the bugle call "retreat" sounded, the call for the flags to be lowered. I was deep into the drug's effects and felt the most intense patriotic feelings for both countries as, first the French, then the American National Anthems were played, as Mother Nature obligingly provided the beauty of a spring sunset as accompaniment. The psychiatrist was finished with me and I returned to Evreux.

My return occasioned a lively debate among the officers. Some wanted me to stay with an offer of an early promotion to E-3, airman second class- two stripes. Others argued about the details of a discharge. It would be honorable, in any case, as I hadn't done anything to merit a bad conduct or dishonorable one. There was the question of reserve duty, regular meetings every two weeks, or Individual Ready Reserve, where I would be called as needed. It was apparent I wasn't completely off the hook. Then there was the question of disability payments like the barber in basic training received.

The final decision was that I would be relieved of active duty, I could be recalled at any time, for any duration, until May 30, 1969, and I would not be eligible for disability. I would receive an Honorable Discharge under Medical Conditions. It was my choice, leave or be promoted. It was not easy, I had been an enthusiastic participant up to the BZ incident and had eagerly awaited my next promotion, but at this time, some three weeks after my three day nightmare, I was still not feeling 100%, and no one had explicitly assured me it wouldn't happen again, so I took the discharge.

I went to an office for my mustering out papers. Here, I had an opportunity to quickly scan the psych report. No mention of psychoactive drugs, no surprise, but the psychiatrist said I was immature and a latent homosexual, so, a de facto scarlet letter. This was the cover story, I was to be gay for my country.

After my papers were typed up, one copy for me, three for the record, an officer came over to the clerk and told him to take two copies over to a different typewriter and add a notation to them, which I later found out were my AFQT scores- Air Force Qualifying Test- the scores that determined my eligibility to join in the first place, and played a role in what jobs I was offered. I wondered why it was only for two copies, why on a different machine, and who would get which copies. My copy had the scores.

Waiting at the transit center on base before going to Paris, a contractor approached me and offered a job paying $400 a week, four times as much as my current monthly pay. The job was in Vietnam. The officer supervising me came over and shooed him away. They had other plans for me. Remembering the warning I had my first day on base, and thinking how many weeks would I last?, I was glad to have help in refusing.

That day before I was to return to the U.S., I was given a room in Paris at the hotel in the Rue de l'Hirondelle, to await the flight the next day from Orly Airport. Before we left for the airport an officer told me to put this opened bottle of French wine in my jacket pocket- a tenth of Burgundy, another smokescreen for my early discharge. Now I was a drunk, too.

At a layover at the airfield in Prestwick, Scotland, I drew duty guarding the diplomatic pouches in the cargo hold. I stood under the C-54, with not a soul in sight, only the building where everyone else had gone in for lunch, a .45 holstered at my side, taking in the windswept beauty of the Caledonian heath. When everyone returned to the plane, I traded the .45 for a box lunch, confident my counterpart had done his job.

We refuelled at Goose Bay or in Newfoundland, and then landed at McGuire in New Jersey, where a mountain stands, named after my Mother's Father's family, and where my paternal grandfather entered this country in 1909.

When we arrived, I found out I would have to stay on base another two weeks, probably to insure no traces would remain of LSD or BZ. It was boring not having any duties after nearly a year of being constantly busy. I spent most of my time in the library. Just before I was to leave, I was given a task. I was taken to a storage room and shown a mountain of potatoes I was supposed to peel. A stool, a small knife, a bucket and this mountain of potatoes. Someone came by and said he would do the job if I would sell him my uniforms. He offered me $60.00, I insisted on $85.00(my summer formal uniform had cost $20). So now there was that much less trace I had been in the service.

If You Can Make It Here….

If clothing was that cheap at the time. Money still had value, it was still backed by silver until 1964. Booze at the PX was even cheaper. I bought a suitcase full of two liter bottles of vodka, my father's favorite beverage, intending it as a gift. I took a bus to New York and checked into a hotel, putting the liquor in a dresser. I then went out to check employment opportunities. Outside a city office, I saw a board listing the job I had in the AF. I inquired but I didn't quite have the certification necessary.

Coming back to the hotel, somebody from the management came to my room and demanded I sell him the booze. Apparently he held the exclusive franchise for untaxed liquor in the neighborhood. It was made clear I had to do business with him, the only question was price. I made a fair profit, he would make more.

I made my way to the N.J. side with my two guitars and a bit of clothing and set out to hitchhike home to St. Louis. A '58 Chevy, black, of the most basic unadorned model, stopped and picked me up. Two guys, short hair, obviously military said they were going to Berkeley, California. I said I was going to St.L. During the ride they tried to talk me into going to Berkeley. Here I was getting a chauffeured ride on the government's dime. Again I was tempted by the offer, but I was looking forward to familiar surroundings where I could pick up the pieces and find my footing. They switched off driving duties, and soon we were on Interstate 70 at the Lucas-Hunt exit a quarter mile from my house. They reluctantly accepted an oatmeal breakfast at the family kitchen. I don't know if they went on to Berkeley, or got different orders.

I was back where I started eleven months ago, but I was not the same person.

Summer of 1962

~

I set up the basement of the family home as my room, with my enlarger and developing trays I had used in high school, a bed, bookshelf and a stereo. A back door gave onto a concrete patio with a wooded area and a creek behind a lawn.

My old '51 Chevrolet, no longer ran. A professional wrestler friend of my father had a car lot, and I purchased from him a 1951 Pontiac convertible, turquoise blue, with lots of chrome- it looked like a piece of Navajo jewelry, and was my transportation as I set about to remake my life.

My first attempt at employment was at Barnes Hospital, on the eastern edge of Forest Park, a leftover from the '04 World's Fair. In the western end was the Art Museum, with circles providing parking, a statue of King Louis lX, and a scenic overlook toward the northern border of the Park, where the Jefferson Memorial stood, a historical museum. And beyond, Delmar Blvd, an entertainment district near a synagogue, Harris Teachers' College and the Scott Joplin house. At the western edge, the City Limit, Washington University, the County Seat, Clayton Mo., and upscale housing. The middle of the park held a Municipal Golf course and the Zoo.

My certification from basic medical training got me an entry level job at Barnes. I didn't seem good at it, and my supervisor thought I should try better things. Through my father's connections I got a job teaching folk guitar at University City's Baton Music. From this beginning, I expanded to other music stores and gave private lessons at people's homes.

Charlie was back after his active duty. He had joined the Army in 1960 and when he was home on leave, he told me he couldn't disclose anything he was doing and resisted any probing. From this, I got an inkling of "need-to-know" secrecy. He came back in '62, did not wear a uniform, but had some mysterious job near McDonnell Aviation that gave him a lot of free time. During this time, and later in '63, I went with him on several occasions where I would observe him ingratiating himself with people who worked at McDonnell . I came to understand he was keeping an eye on people suspected of stealing from the company, or of passing information to the Eastern Bloc. Once we went to ask about an apartment for rent (none had been advertised) just so he could see inside for a moment. He was also involved with a Baha'i group- keeping an eye on them?

Charlie and I had been friends since junior high, both our dads worked nights, and Charles senior, who was a union projectionist, and was a veteran of the union struggles in the thirties, was often available during the day to pass on his considerable mechanical knowledge. Charlie replaced a starter motor in a '39 Pontiac coupe which had replaced my convertible when the driveshaft fell out of the '51. He also rebuilt the transmission of a late forties Ford Anglia in his garage, and helped me convert my first car, a Ford Model A from a coupe to a roadster. We probably put less than 35 miles on that and it ended up in a junkyard.

Charlie and Mike's Excellent Adventure

When I was 16, during the summer, Charlie had this idea we should run away to Corning, Ark., where his grandmother lived. Not much reflection on my part. It was all planned out… his mother was out of the house, in the dark of the evening. He had called a cab to pick us up on a dead-end street up a short wooded path behind his house. With perfect timing, we reached the street when the taxi did and we were on our way to the station. He had me dressed in one of his dad's coats and a Fedora. He had two tickets already purchased. I was to be his father and somewhat ill. We boarded the train, I sat at the window seat, he handled the transactions with the conductor. When needed I pretended to be asleep, with the hat over my eyes.

There was a short layover, fifteen minutes, and in order to preserve our cover, we got off and go to a cafe out of sight of the train. We got back on with a minute to spare and soon we had reached our objective. Charlie knew a guy with a little radio station, and we stayed in the nightwatchman's trailer our first night. We then got a room behind one of the few restaurants downtown. Late at night, after closing, the smell of poor quality hemp wafted into our room.

His grandmother agreed to take us in and not inform her daughter, Charlie's mother. The day began with Granny drawing water from a pump in the kitchen, stoking the fire in the large black cookstove, the fire having been started before we awoke, and her preparations of a hearty country breakfast. There was cornbread, biscuits, beans with bits of bacon, greens, hominy, and eggs. Then the adults got ready for whatever work was available that day. Cotton was ready to pick, and in this area only whites are selected for this work, which will yield a cash payment at the end of the day. Not a great way to make a living, what with the way the dry bolls cut your hands up, but it's just one of the things you do there along with hunting and fishing, tending your garden plot, and taking payments from the "timber hawk," who pays a fee to cut trees on your plot, usually separated by some distance from your planted land. At night everyone gathered around the pot-bellied stove in the middle of the living room. Its intense heat balances the cold of the minimally insulated house until we retired under a mound of quilts and comforters.

Charlie knew a mechanic in town with a quonset hut for a garage. I bought an old Plymouth from him. The brakes weren't really good and at the town's only stoplight I bump into the driver in front of us, no damage, and Chuck's skills at fast talking and citing all the people he knows in town keep us out of trouble.

The town retained an old southern feel, there was, instead of a photographer, a man who would cut cameos, silhouettes in a booth that would cast the shadow of your profile for the purpose. It was here I first heard Johnny Cash singing "I Walk the Line", and the Everly Brothers.

Catholics were not welcome in the town. Not only were they still fighting the war of the Southern Secession, they were still fighting the English Civil War. Sunday came and Charlie and I were going to Mass. There was a building downtown with a side entrance and stairs to an "upper room". People would filter in singly, while the good townsfolk were at their churches away from the downtown area. After the congregation was assembled the priest came in, in civilian clothes, put on a cassock, withdrew the missal necessities from a small case and said the Mass. After, he stayed a while to attend to pastoral duties as we all left the place, again one or two at a time. The Kabuki served to keep it out of their faces.

In a scene out of The Last Picture Show, Charlie and I take two local girls on a double date. Necking ensued. Later, my date's father, known to be abusive, raped her as punishment for dating an out-of-towner, apparently. Within a week, the town had decided to run him out.

By this time our parents knew where we were, and we took the train home. I'm not sure, but I think this was Charlie's plan all along. It was not the only instance of him rescuing a damsel in distress. My date eventually moved out of the household and went to live in Montana.

Some Notes on St. Louis Geography

St.Louis was founded on Spanish territory, wrested from Osage people, on a hill situated in a convex bend of the Mississippi River, a few miles south of its confluence with the Missouri. The old town was bounded on the west by 12th St., on which the French-style Hotel de Ville was built, the area having been ceded to the French. The old cathedral had been built not far from the river. As the city grew after the Louisiana Purchase, a series of streets radiated north, northwest, west and southwest from this center, streets that would form the boundaries of my youth. On the west- northwest border of the city, Page Blvd. brought a traveler to the early suburb of Wellston.

My mother's grandfather Caleb's farm outside St.Louis encompassed an area that would later be developed into housing, Wellston High School, and the Wagner Electric Factory. A creek running south split the farm, and by the time I was born, provided water for the factory and was hopelessly polluted. My grandparents' house stood on a remnant of that land and still had a barn in the back yard. My parents' house was a couple of doors east, and had two wells from the farm that had been filled in with gravel. The street was Plymouth Ave., which ran east and west. Our houses were on the north side of the block, so the view off the front porch was south. I grew up associating south with "up" and north with "down". The sun, therefore set to the right, and downtown was to the left. My mental map was upside down. When we later moved to a house my uncle Jim built, the back porch provided the southern view and my old neighborhood was therefore up Lucas-Hunt road several miles. Our house and the house of Charlie's family were on the grandly named Roland Blvd. At the bottom of the street the two best houses stood, one owned by a prominent Anglo family, and the other by members of the local Italian aristocracy. The father drove a new black Cadillac, and the son a new black

Chevrolet convertible. The rest of the neighborhood was comfortable working class- lots of union members.

West, up the street, across Lucas-Hunt Rd. was Normandy, a township named by the many veterans of that campaign who had built there. There was a Norman style tower that stood on one entrance to the place. Many of the houses were substantial and reflected a California Spanish style, others were English Baronial.

From where I lived, Lucas-Hunt Rd. was the way to get anywhere. North to the freeway offering access east to downtown and west to the airport, continuing north past the freeway passing the cemetery where my grandparents would be buried, to the new Northland Shopping center.

Going south on the road got you to Natural Bridge Rd, left there, another way to get downtown. At one time the road crossed a southwestern style natural bridge, which was subsequently covered by undergrounding the stream that had formed it. Going right at Natural Bridge led two places, the intersection with Florissant Rd. which veered north a bit to reach Ferguson Mo. and the left fork which continued on past Normandy Junior High, the Golf Course being redeveloped as the St.Louis Campus of the University of Missouri, and on to the Eero Saarinen designed Airport, Lambert Field.

Ferguson had been the site of an early conflict I had with a type of person I considered completely irrational, and devoid of any common sense. I was at the cashier's window in the local department store that served as the motor vehicle department's local office. I merely wished to pay the fee required to get my driver's license. The white woman at the counter, for some reason didn't want to serve me, but didn't have the guts to say why. Fortunately the majority of bystanders took my side and I walked away with my license. The reader will note the presence of Ferguson in the news recently.

Past the airport, Lindbergh Blvd. curved to the south, a western beltway that was at the time part of Route 66. It continued on to the suburb of Kirkwood then turned west to become the fabled

"Mother Road"- " you'll see Amarillo, Gallup, New Mexico. Flagstaff Arizona, don't forget Winona.."

Continuing south on Lucas-Hunt would get you to St.Charles Rock Rd. with Normandy High School at the intersection, site of many formative experiences, one of the most positive of which was mechanical drawing class- drafting, where I learned to turn three dimensions into two, while not leaving too much of a mess in pencil lead.

In my time the road which led to St.Charles, across the Missouri River, had become paved. Going west, there were auto speed shops, the Harley dealer and a big box store not subject to St .Louis regulations where an insecure thirteen year old could buy an Italian stiletto, a brand of switchblade which resided in my jacket pocket throughout my entire thirteenth year, and was never once brandished. When I turned fourteen it was gone.

Turning left at that intersection led to Wellston, Forest Park, Olive St. and eventually, Union Station, Kiel Auditorium and downtown.

These were the boundaries of my life there.

That summer I taught lessons in the afternoon, and evenings went down to "Gaslight Square," the new entertainment district to listen to other musicians and do mostly unpaid gigs. A hurricane had, a few years earlier, devastated a two block area of Olive St, near Forest Park and the most distinguished hotel in town, the Chase Park-Plaza. Olive St. had been rebuilt with a wide sidewalk and a lot of clubs and restaurants. Before I went into the service, I had worked at the "Crystal Palace," a theater restaurant, mostly inventorying liquor. I moved into a building housing mostly students at St. Louis University (Catholic, a basketball powerhouse), and denizens of the entertainment district.

I started freshman year at the new University of Missouri campus, next door to my old Junior High. I took philosophy, sociology, English lit, and a math course.

One day, I was called to the office and told that people from the State Department wanted to interview me. Naturally I was flattered, but didn't know what to think. The interview went well until

I was asked what I would do if I received orders to do something that went against my principles. I felt the answer they were looking for was something like " my country, right or wrong". I hesitated, remembering seeing the TV presentation of Judgment at Nuremburg, and a lecture we had in the Air Force about illegal orders and that we couldn't just say we were following orders. I said "I would talk to my supervisor." They said "what if you couldn't reach him?", and I said I would follow my conscience. At this point they said the interview was over. I pressed them about what I should have said and they finally said "you would have to resign." For good or ill, I would not walk through the front door at State.

After I left the room, two tough-looking guys took me aside, and said I have to forget about everything that happened that day and in the service. If you want someone to forget something, you shouldn't constantly remind them of the fact, but they continued to threaten me and press me on it. I used my knowledge of "Omerta," the Italian code of secrecy, and just repeatedly insisted I didn't know what they were talking about. They continued to press, I continued to deny, and they were satisfied and let me go.

One day, some of the more popular kids I had known in high school, who didn't want much to do with me back then, invited me to a party at one of the better houses on my street. I was flattered to be included within that circle. For entertainment, a hypnotist exercised his art, but when my turn came came, he was unable to hypnotize me. My training kicked in and I was deliberately resistant. It may have been a test. I haven't been hypnotized since that day in basic training.

I finished out the semester, and resolved to go to Boston to look up a fellow Airman who had also been discharged early. He was one of the privileged diplomatic sons from my days at Gunter, and had been at Evreux. Somehow, I felt I might be able to put some of the pieces of my life back together. Between the secrets, the threats and the orders to forget…

The Midnight Ride of Michael Paul Revere

~

So I find myself back on the road, bus to East St.Louis, US 40 across Illinois, crossing the tracks that mark the Indiana border, and finally reaching the western suburbs of Indianapolis. It's snowing and late. A car with four guys my age stops and picks me up and offers me a few square feet on their floor to stay the night. I gratefully accept and fall into a peaceful sleep, with the wind whistling outside. During the middle of the night they get up, and I hear conversation that suggests they are gay and are about to have some fun, They invite me to join them, but are OK with my refusal. I keep my eyes closed and eventually fall back asleep. In the morning one drives me to the Greyhound station, where I buy a ticket to the next town, Greenfield, which happened to be where some of my relatives were from. The guys may have been Defense

Intelligence and the "party" staged, or it was as it had seemed to be. I was now in a world where sometimes things were not always as they seemed.

Stopping in a cafe for a cheap breakfast of coffee, toast, butter and jam, I was surrounded by friendly people who wished me a safe journey as I resumed hitching to Boston. I got a series of rides and eventually followed the western hills of Massachusetts down to the sea. I found the address I had been given, but nobody was home. My friend from Gunter and Evreux had said his father was an architect. I looked up Architects with his last name, and phoned him at about three in the morning. When the person answered but wasn't my friend's dad, I received a fair amount of grief from him. After spending the night in Boston's North station, where I managed to get my suitcase stolen, I set back out on the road, ending up at my uncle's house in Schenectady, NY, where I spent a couple of days visiting before finally meeting my friend in the Boston Suburb of Dorchester. He was half Jamaican, so let me call him "Bob".

I went to the Employment Development Department, and the interviewer tried to talk me into going to Hawaii, saying jobs were scarce locally. I had neither the inclination nor the means to make such a journey, and when I balked at his suggestion he called someone, and I was offered a job at Mosher Music Co., an instrument, sheet music, and record store across from Boston Common.

Bob took me to a cafe filled with a better class of people than I was used to in St. Louis, professionals and academics. He introduced me to an older man who seemed to be interested in making sure I followed certain ethical standards, who gave me a book that illustrated themes of social responsibility and personal ethics. He also suggested that I take up photography, already a hobby of mine, but cautioned me not to buy a Minox(the Latvian invented "spy" camera.) When I later visited a camera store, that is exactly what I bought. I took pictures of the back alleys I traversed in the course of one my duties, delivering to the

Post Office the packages I had wrapped for the music store's mail order business.

One evening Bob and I rode buses around while we had a heart to heart talk- he wasn't happy I wasn't interested in a sexual relationship- and I was soon looking for new roommates. Through an ad in the paper, I found a place on Commonwealth Ave. in Brighton, across the Avenue from Brookline. To me this was the height of sophistication and class. Brookline, where the President lived when he wasn't at the White House or Hyannis.

My new roommates were Vern from Minnesota and a guy from Oklahoma. Both were music majors at B.U. I learned a lot about classical music from Vern, to supplement what I knew about jazz from my father. I was still used to being busy and soon had two more jobs, one at the Community Music Center, which gave cost-free lessons for poorer kids. The director was a kindly white-haired man who spent some time encouraging my musical development. The other was job teaching a night class in folk guitar at the Boston Center for Adult

Education. After I talked about my experience and qualifications, I was questioned about racial matters in St.Louis. I kept trying to convince them I wasn't a racist, but they kept trying to find out if I had been discriminated against. Identity seems to depend not only on who you think you are, but also on what people perceive you to be. I got the job, in any case. A student in the class took a shine to me, and we become friends, continuing past the end of the class. I will see her again, in St.Louis and later in San Francisco. We were scheduled to go to Hyannis where the Kennedy compound was one weekend, but the trip was cancelled.

I bought an old Chevrolet for $25 from a gas station that had seized it on a mechanics lien. My Oklahoma roommate had just got new plates for his car and I put on his old ones. In the urban hubbub around B.U. no registration, no insurance, only the receipt from the gas station- no problem.

Vern and I decide to drive up to Kennebunkport,Maine. Nice drive up at night. We parked by the rockbound shore where we could look across the water to the Bush compound. We start back on Saturday afternoon, on Highway 1. I was pulled over for an expired plate. The trooper was suspicious but let Vern go. I spent the next two nights in the Ogunquit jail, the Judge being unavailable until Monday morning. I used my phone call to call my father and he and the jailor have a sergeant to sergeant talk, which lessened the pressure on me. Monday, the Judge talked to my employer, I paid a ten dollar fine and took the bus back to Boston. The State of Maine kept my car.

In our apartment we had an agreement that if one of us has a lady friend over, the others have to stay away until midnight. Kathryn visited and we listened to music and had as much fun you can have without risking disease or procreation.

As a Missouri folk musician I'm somewhat exotic in town. I can play the basics on 5-string banjo, which I give lessons on, and I was at the Folklore Center in Cambridge, admiring its collection of instruments and someone asked me to broadcast with a small group on a student radio station. There was talk of Joan Baez who had played and sang the week before. I got nervous as we started playing, but it passes for authentic old time music.

Having pretty much had the New England experience, I make my way back home. Not much later, I was at the Northland shopping center, looking at radios, when JFK came on one of their TVs and made an announcement about Soviet missiles in Cuba. Everyone there had gathered around discussing the news.

Music and Mexico

~

I moved back into the building at Gaslight Square. It was a three story building with wrought iron balconies on the upper floors, giving it a kind of New Orleans look, which fit in with the club up the street featuring a Dixieland Band. I got another used '51 Chevy, which I spray painted white in my parents' driveway. I resumed my musical activities; I had the use of a practice room at Washington University, teaching faculty members' kids, and taught at a music store in Valley Park, Mo., which had a new Chrysler plant.

From the time I was relieved of active duty, during '62 and during the time I was in Boston, there were no surprise doses of LSD, but one day in '63, I received a phone call telling me to go to a particular address. I tried to refuse and asked who was calling. I was told never mind who it was, and if I didn't comply, I would be hauled before a grand jury. I was a bit puzzled since I wasn't involved in anything criminal. I guessed it was my old "friends" calling me up based on my period of military obligation. I went, and someone there questioned; "why did you bring him here?", a superior quashed his concern, and offered me something to drink, and we went outside to the back yard. The gathering dusk and the onset of a dose gave me an appreciation of the natural beauty of the evening. We talked, about nothing consequential, and I was allowed to leave.

I met a W.U. graduate student, a baritone. We rehearsed Elizabethan songs by John Dowland and others. I accompanied him on guitar for his master's degree recital. I also joined the St.Louis Early Music Society, singing bass and playing lute music on guitar.The practices and recitals were held in a church near Concordia College, a Lutheran School. I would sometimes attend Lutheran services to hear Bach's choral music.

Saarinen had another project in St.Louis, the Gateway Arch. A documentary film with a symphonic soundtrack had been shot to display there, and the local educational TV station wanted to air a version with a different score. I was given the opportunity to record classical guitar pieces for the TV version, which formed the score except for a saloon sequence using a player piano.

I met another Michael at my building and was introduced to a different circle of friends. He was also studying classical guitar, and we became friendly rivals. At one point four of us went to Mexico in my white Chevy; Michael, Sue Ann , Jerry, and myself. Looking at the map we picked a spot on the gulf coast, La Pesca, in Tamaulipas State. We started down Route 66, I did the driving. In Oklahoma City, we were delayed by the need for an engine repair. Back on our way in Texas, we noticed a marked change as we descended into the Rio Grande valley.

We crossed the border and loaded up on cheap Mexican cigarettes and marveled at the prices of beer and liquor. We got to Monterrey where we got a large room in a tile roofed traditional looking place. Sue Ann was Michael's girlfriend, but on this trip they agreed to forego intimacy to preserve group amity.

We continued on to Ciudad Victoria where the road, which had been blacktop,became loose rock. Buildings and walls had signs painted urging one to vote for Gustavo Diaz Ordaz.

We struck out for the coast, and the road became dirt through undeveloped mountain land, beautiful in its own way. The only way you could tell if you were on track was by looking for signs advertising Coca-Cola. At one point we erected a little shrine of coke bottles in thanks. We drove along and came to a ranch. We were informed that the road was on private property. We apologized but then were invited to lunch by the proprietor, then sent on our way with a "vaya con Dios". That night, I was driving and saw what looked like two headlights, seemingly far away. Only when the '58 Chevy pickup was a couple car lengths away could I see that it was one side of dual headlights; the driver's side lights were burned out: a quick stop by both drivers.

We saw Soto la Marina in the morning. There was a festive mood, this being Saturday, music playing from the bandstand, and a crowd in the plaza. Dirt roads and a stream to cross, we must outmaneuver a bull to get into town. We draw a certain curiosity but Michael's fast talking allays suspicions. We buy supplies in one of the stores on a secondary dirt plaza and leave town in the afternoon. We were on flat marshy land now.

We approach La Pesca to disappointment. It's a "contra costa", a strip of water, then an offshore peninsula that hides the ocean from our view. No beach. No gas station. The only business in La Pesca is a small airstrip, and a hangar populated by Americans unhappy to see us. Did we stumble onto something outside our "need to know"? Once again Michael allays their concerns and they give us some aviation gas- no service stations within 80 miles. We resume the journey inland to the southwest toward Ciudad Mante, which we could have reached directly by paved roads from Victoria, though we would not have seen that beautifully wild undeveloped area. At Mante a meal of steak, "frijoles refritos" and tortillas. A well-paved road brought us to Tampico in Veracruz state, and its seaside twin, the oil refinery town of ciudad Madera, a prosperous place filled with young workers.

We had mounds of boiled shrimp and went to the beach. That afternoon, I fell asleep on the sand and was seriously burned on my back and the backs of my legs. As it happened, I was the lucky one; I had refused to eat anything not thoroughly cooked the whole trip, while my companions eagerly ate the lettuce and tomatoes. They came down with "Montezuma's revenge" and I had to be the "go-fer" while they suffered in our hotel.

While in Tampico, we needed a couple of tires and got two adequate ones for about four dollars.

Our return trip was uneventful, passing U.S. Customs with no problem, but had to pay Texas

Customs for the small amount of liquor we brought back.

After we got back, I sold the Chevy to a friend of ours for fifteen dollars. He couldn't get it to start. It had served us faithfully, but now was put out to pasture.

For any consumer item and particularly musical instruments and cameras, Chicago offered a wider selection. A man called Jim Sherry, with a partner, a Mr. Brener, operated a Flamenco dancing studio and a business importing guitars from Spain, in an outer suburb. Michael and I went to find better instruments than those we had. The mainstay of "Antigua Casa Sherry-Brener" were guitars made by Jose Ramirez in Madrid, and Japanese made instruments with labels that would lead the customer to believe they were also made in Spain. Andres Segovia played a Ramirez in concert, and had a side income selling them to students. Segovia also figured in Sherry's marketing. Segovia gave master classes in Berkeley Ca. in '64 attracting classical guitar students there.

Michael selected a Ramirez and I bought a Francisco Fernandez instrument. Years later, I looked for the shop in Madrid, but even the street address was nonexistent. Fortunately my japanese-made instrument exemplified that county's artistic quality and served me well for many years.

Downtown Chicago had music stored with very fine violins, but not much selection of guitars. There were, however, several large camera stores in the "Loop" serving the newspaper industry. I had bought a Speed Graphic camera, the kind Weegee used, the big boxes that now define the era in films about the forties.

One time I wanted to look for a Leica 35mm camera in the windy city, and set out hitching north on 66. My second ride took me as far as a restaurant complex some forty miles south of Chicago. There was no signage. It was apparently not open to the public. the large parking lot was full of big fancy cars. One knew it was a meeting place for those on the inside of a non- governmental organization of men concerned with honor and secrecy. So my ride had left me off in front of the place and entered.

I was standing out there with my thumb out and someone who had been tasked with finding out who it was hitching outside their private club, picked me up and drove me north to town, questioning me, finding out I was a former airman and on my way to buy a camera. I think he was relieved to find I had no ulterior motives. I had pointed out it was a member of his fraternity that had dropped me there.

I got my camera, a Nicca, a Japanese copy but with an authentic Leica Summar 50mm lens. I was impressed by the sharpness of the photos.

That autumn, I tried another semester at U. of Mo. I took a poetry class, sociology, which I found eye opening, and a history class.

Near campus was an early franchise of McDonald's With 19-cent burgers, and a coffee house with a stage where I heard a new group, Peter, Paul, and Mary.

Earlier that year, I had been going out to the Eero Saarinen designed airport, Lambert field. There was a free parking lot across a small road from the main terminal entrance, where I would park and go watch planes and practice photography. There was a small room with arcade games. I regularly played one that had a rifle stock and a mirrored surface that displayed a grid of small tennis type balls. You aim at a ball and if you hit it at just the right spot, it bounces out of its socket and you get points. I got fairly good at this.

One day, I noticed it had a different rifle stock. I was playing and two guys came up and offered me a cocktail, which I drank. After a while I got the usual feelings and the lights on all the games looked remarkable. I hit the spot on one of the balls and all of them start bouncing wildly. I left the room, and the two guys told me to say something about his wife to this older guy as he came near. He was tall, white-haired and becomes agitated. I got in trouble with his security people, but my "security people" say it's OK, I'm only drunk. But the impression is made. I later recognise him as Clark Clifford, and my guys have my prints on a rifle stock.

Not long after, I heard JFK would be coming to speak at the airport. I came a little early and parked at the free lot. Across the street was a small crowd. Some are well-dressed, looking like FBI or Secret Service types. A small bunch, not well-dressed, of four or five are milling about in a circle carrying signs. The signs say among other things, "Fair Play For Cuba" One well dressed guy, not part of the group marching told me, as I was photographing them, that I shouldn't just take pictures, I should get involved. Did he say the code word, as in The Manchurian Candidate? I joined in. another of the well-dressed said Kennedy was actually going to speak at this other area. We are shuttled over there, but eventually nothing happens. We had been given the bum's rush. The marchers say come with us to this nearby bar. It's closed but they have the key. They may have convinced me to give them the film. I never had a shot of the "demonstrators".

They treat me as an insider. There was no unified command, one bunch tells me to forget, another pulls me back in. They ask me to study the"Black Liberation Movement." The Fair Play for Cuba Committee has asked me to keep an eye on the Nation of Islam.

Later that year Kathryn called me from the airport, and I met her there. She was moving to San Francisco, and gave me an address in case I come out there.

Gaslight Square and that Day in November

By fall of '63, the square was fairly bustling. Starting from the western end on Olive St., where you could usually still find parking, on the south side of the street is a fancy restaurant and some offices in what used to be the "Music and Arts Building." This used to be studios in the thirties and forties. Around the corner on Boyle St. was a bar featuring guitar music,and then a cafe with a stage, the"Laughing Buddha". In the foyer a sale in the cigarette machine would produce the sound of the buddha laughing, presumably no irony intended. I sat in the "green room" with Judy Collins as she warmed up for a set upstairs. Peter, Paul, and Mary whose sessions out by the University were sparsely attended, drew a large crowd at the Buddha, and the espresso machines were busy that night.

Next door a larger venue was soon to open. Across the street a progressive church offered social services and had a coffee shop with a stage. Next door an Irish bar decorated with mirrors advertising Irish Whiskeys and Beers. Behind the bar, there were two recessed pots, one with today's soup and one with tomorrow's, the most food for your money around. Someone suggested that I keep current with Irish politics, I looked Irish, but being Welsh, I had no clue, and I had plenty to do keeping up with African-American news.

Around the corner, on Olive St. the largest business was the "Crystal Palace" a theater night club. At the end of the street, around the corner, was the Dark Side, a jazz club and the only substantially integrated venue.

Over on the north side of the street, was my building and behind it, a folk club, Jacks or Better, owned by a guy named Jack, which blared recorded music by the Irish Rovers, but actually featured folk acts doing their version of the Kingston Trio. My job there was to play boring classical guitar music between sets to clear the tables. Next up the street was a restaurant with red velvet wallpaper made to look like a Big Easy bordello, then a larger venue for Yardbirds clones, a couple of antique stores with a brick paved area in front, where I sometimes sold original paintings, some were mine and others were from art students I had known in high school. The dixieland club was next, followed by a blues-oriented club, a gay bar, and a Mexican restaurant with a stage that provided another venue for us guitarists. Across the street, another blues and bluegrass place where I sometimes played bass. Several of us cobbled together a living in this way.

Some people were aware I had a little knowledge of psychedelics. The only way they could have known this was insider information, since I downplayed anything about my military experience. A group of people encouraged me to process morning glory seeds for an all night party they wanted to have. About twenty attended. Dylan's album with "Positively Fourth Street", and a song by the Byrds played on the phono. Capsules were taken and people mostly tripped in silence. Not long after, at a small session including Jerry and Michael, we were playing the United Nations World Music Series album of Tibetan chants. It was late and I was tired. The sun was beginning to rise. Someone produced some pot. I dutifully followed the orders that had never been revoked and smoked some, and went into a bedroom to lie down. Lying on my back looking up at the white ceiling, I saw colored lights caused by cars passing outside, projected camera obscura style above me. These coalesced into gears and wheels within wheels swirling in pastel colors until they suddenly stopped. I felt an intense calm or tiredness, then I had the sensation of ascending in pure light into what seemed to be an infinitely large room without windows and seemingly without bounds. I was suddenly back on the bed, got up and tried to explain to the others what I had experienced, but was inarticulate.

One day I was feeling particularly pessimistic about racial polarization in the U.S. I reasoned I had to get out of the American perspective. I needed a different viewpoint. The Soviet view was useless, but what about America's WWII enemies? Germany wouldn't do, but what about Japan? I had noticed a Japanese import store near Union Station. I drove there and checked out the ceramics, the fabrics, then I spent some time perusing their bookshelf. Most of the material was actually about the Japanese-American experience. I had thought I knew a lot about our country's racial and immigrant history, but now I was face to face with a story I knew next to nothing about. I read about the camps and the Japanese-American soldiers in the Pacific and European theaters during the war. I looked through a cookbook put out by the Japanese American Citizens League, with recipes that preserved their culture with what was available here. The proprietor came over unexpectedly and asked if I had a radio in my car I could listen to. President John Fitzgerald Kennedy, The man I had sworn my allegiance to had just been shot in Dallas.

I went out to my car and listened to Walter Cronkite, the most respected newsman of those times, announce that Kennedy had been shot and it was not known if he had survived. Throughout the day details would dribble in, information and disinformation.

I finished out my semester at the U. amid the shock and dismay of my fellow students. Most people I knew had fully supported Kennedy presidency. My father had played piano for many campaign fundraisers at Irish bars in town, and when Frank Sinatra announced his support, we Catholics felt we we would finally be fully respected by the Anglo establishment. School ended, with only final exams to go, but I felt compelled to take up Kathryn's invitation to visit her in San Francisco.

On a cold day in Missouri I set out to hitch there. First I drove my car to Pacific Mo., parked it in a large lot where I was fairly sure it would not attract attention, and put out my thumb on highway 66. My first stop got me a ride to a truck stop I had passed before. It was full of drivers who looked like the people I had worked with on my uncle's construction projects. The difference this time was that I had a beard, and they jeered me. Down the road, I found a barber shop and had it sheared off. This might not have been necessary, since my next ride was from a musician moving to Arizona. The V.W. bug he drove had a Hammond Organ stuffed in the back seat, with my guitar, the one I taught my folk guitar class in Boston with. My ride took me as far as Kingman, I think, then I got a ride to Barstow, next Mojave and eventually on a spur of 66, U.S. 466, which now is California 46. A ride up the west side of the central valley on 33, a rural highway through sandy desert land. This was before the aqueduct and Interstate 5 were built.

I met Kathryn and her roommates in the Marina district of S.F., just off Lombard street. I checked out employment options but didn't find anything appropriate and wonder at the high gas prices. 38 cents a gallon vs. the 25 cents I've seen across the country. She and her mates are Navy nurses, a fact that comes as a surprise to me. She traded a Mexican guitar that doesn't quite work right for mine, and I use the workshop in the building to repair hers. I decided I needed to get back to my jobs in St. Louis and hitched back. I was duly impressed by the winter temperatures in the Bay Area, and got the idea I may move there eventually, fulfilling the suggestion of the guys in the '58 Chevy back in '62.

Back out on the road, thumbing. I was insulted by right wing locals outside Barstow, I pointed out I'm leaving California so they should be happy. I got a ride to Las Vegas with an Air Force officer and his family, among friends now. I ran out of funds in Henderson, and called my dad for help. He's not too happy, but he has a musician friend in Vegas where I stayed a day or two, then on to Flagstaff, with its forested beauty. In Santa Rosa, New Mexico, a long wait for a ride at sunset by the river, planted the seed for my love of the area. A few more rides and back in St.Louis.

I picked up my musical efforts through the winter and spring, still living at Gaslight Square. A black blues singer, a big guy, and a big talent, became part of our circle. He and I started attending meetings of African-Americans in the neighborhood concerned about their mistreatment. People were concerned by my presence at these meetings where a range of opinions were shared but the singer presented me as his "mascot."

Parallel to Olive St., to the north, is a black business district, Washington St., and down that street, near my apartment, is the office of the black Musicians' Union. My father used to take me there, where in a lounge in the basement you could socialize, get your shoes shined, and if you were known, you could get a little box of marijuana. The white Musician's Union, Local 2, in the southern part of the city, was in a small industrial park, with two identical cinder block buildings. In one, the Union, in the other, one of the two places in the U.S. where cocaine was extracted from coca leaves for sale to dentists. The denatured leaves were then shipped to Atlanta where an extract was made to flavor Coca-Cola. You couldn't go into the cocaine "factory", but there was a dentist nearby where you could schedule a procedure, pay for it, and receive, in lieu of treatment, a small amount of the drug.

In St. Louis, that is, the city, not the county, the southern part is working class white, and the near north of the city is the black neighborhood. A perusal of apartment ads in the newspaper reflects a big differential between rents in the north and the south and west.

The city was surrounded by a multiplicity of townships under St.Louis County jurisdiction. The major streets crossed these boundaries, but the neighborhood streets did not connect with the streets of the city. There were no barriers, but there was an unpaved area, a visible color line. One evening Charlie said we were going to meet two black women he knew and we crossed one of these gaps. White perceptions of black women being what they were I expected a party. It turned out the music they played on their phonograph was a record by Louis Farrakhan. It started with an Arabic sounding melody he played on the violin. It was beautiful and made me wish I could play that instrument. The song that followed was an exposition of Black Muslim, Nation of Islam, philosophy. Not the evening I was expecting, but when I later attended their meeting, I had some preparation.

I read a small ad in the Post-Dispatch for a Nation of Islam meeting around 1PM at Kiel Auditorium. It was in late spring. I took my camera. I was always thinking I could get something to sell to the newspapers, but it never happened. Strangely, as I was approaching the steps to the Auditorium, nobody was in the streets. Instead of the other attendees I expected to see and photograph, I was the only one coming in. Only two large black men for security, no one else in the lobby. They say they have to take my camera, I have to expose the film.I ask" Why, I haven't even taken any pictures yet?" They didn't look like they would be receptive to argument. They could have been FBI plants, so I couldn't argue with them either. I went into the seating area. The place was completely filled. I figured later the audience had been bused in earlier. The men were all in front, in dark suits. The women fully covered except their faces, all in white. There was a two row area on the right side, at the end of the men's section, reserved for the press, and for other whites. I took the last seat, on the aisle. On the auditorium floor, in front of the stage, was an arc of well muscled men, the "Fruit of Islam", and on the stage, which was decorated in some sort of "oriental" theme, was a lectern, stage left. The first speaker was Wallace Muhammad, Elijah's son. He talked about the Chicago Mosque, and how former prisoners had turned their lives around, and became the Fruit of Islam, and how the organization had its own schools. Farrakhan was there, he was introduced, but didn't speak.

Malcolm X spoke next. In contrast to Wallace Muhammad, he was a charismatic orator. He spoke on the themes we are now familiar with. I have never been able to find a transcript of that St.Louis talk. Malcolm had resigned from the N.O.I. in March, and had his own Mosques independent of the organisation. He had just gotten back from the Hajj, the annual Muslim pilgrimage. He gave an anti-capitalist speech, usually not part of N.O.I. rhetoric, talked about the African liberation movements and how the deck was stacked against the "so-called Negro". He spoke about an uprising against segregation, including well meaning whites. (N.O.I. was OK with segregation- they wanted a separate nation for blacks, and eschewed any cooperation with whites.) Malcolm said a few words about mixed race people, looking over to where I was sitting as he did so.

Elijah Muhammad spoke next. I had read a few copies of their newspaper Muhammad Speaks. His theology was nothing like what I read of in "The Religions of Man" by Huston Smith. He literally claimed to be God, and that was all you needed to know. He spoke about some of his tenets; that white men were devils, that the "Asiatic black man" was the superior race, and that white people were developed by a mad scientist. His sermon seemed to me a mirror image of what a virulently racist white preacher might say.

After hearing folks like my grandfather being insulted, I got up and walked out as he was talking. I think E.M. stopped speaking momentarily at that point. If anyone on that stage had paid any attention to me, he would have noted an enthusiasm for Malcolm's remarks and a disdain for Elijah Muhammad's.

Concerned about getting my camera back, I went back to the lobby and wangled my camera back from the security people I had encountered before.

Back to Paris

~

Kennedy's death broke something. It seems we were just going through the motions afterwards. The chain of command now had Lyndon B. Johnson at the top. Five more years until I would be off the hook. I was basically able to pursue my own goals with relative freedom, and was happy to cooperate with my circle of friends in the light duties our "reserve" obligations imposed. We had no regular meetings, and no two week call-ups that regular reservists had. Just shut up about certain things and everything was hunky-dory.

Charlie confided in me stories of his duty in Korea. He taught me the the system of syllabic characters, which I have since forgotten. We made a few trips to the Missouri School of Mining and Metallurgy, where he had some Korean exchange students he checked up on. He told me of his work in country with the N.S.A., where he drove a Dodge Van, the same vehicle we had used in Evreux, filled with electronic gadgetry to monitor communications coming from the north. He shared stories of his duty station in Ui Jang-Bu, with details of the people's houses, with the underfloor heating they used, and how they would put aside leftover rice mixed with water, which would ferment in a few weeks into rice wine. He had a bucket of it stashed in his closet. We were both influenced by the time we had spent overseas with Uncle Sam.

A folk singing club started at Washington University, and I attended the first meetings. Individuals and small groups would do what they had practiced and sometimes we sang as a group. Soon, a split developed. There was a contingent that only wanted to do American music, basically bluegrass. This group was only guys, but the group that wanted to do international stuff was mixed, and one of the women convinced me to join that offshoot. There was Israeli material and a lot of Latin American stuff which fit with my classical guitar emphasis. We continued to meet on campus and the bluegrass boys met elsewhere, but we sometimes crossed paths.

I started to plan a trip to France, to perfect my Classical playing. I wrote to the State Department Office for Study Abroad, and received a brochure with addresses of several schools, the contents of which I shared with the other Michael, who would later attend the Schola Cantorum there in Paris. I wrote to a few, and in a change from my intent, The Ecole Mozart offered me a teaching job. An offer I couldn't refuse. I was to start in November, but I had already booked passage on the Cunard Ship Mauretania for late september. It was going to be a squeeze, financially.

Hitchhiking through eastern Indiana, I was at a rest stop, just a wide spot in the road, when a car stopped with an older man and three younger guys. They talked me into visiting a State Park nearby, with a waterfall. After we were there admiring the view, one of them offered a joint which I was, by a standing military order, obliged to accept. Two of them were with me while the older man and one other were in the car at some remove. Suddenly one of them said that they were going to kill me. Am I with Klan members? I turn over in my mind the possibility of running, but that would just turn it into a hunt for them, And I didn't know the territory. I had been ready to die for my country, but this seemed a useless way to do it. I was wondering what to do, and the thought came to me clearly. I said "all right, you just better be damn sure you're authorized!" It came out sounding like a something from a drill instructor, which seems to catch them off guard. They told me to wait while they confer with the people in the car. I stand at attention, as rigidly as I would for any officer, as they check with the older guy and then come back and say they're bringing me back to the highway. On the way back, the younger guys plead with me not to turn them in, undercutting their pleas with threats. Either these good old boys could tell I belonged to Uncle Sam, or they were in on the game and testing me.

Not a half hour later, my friends from St.Louis pull up in a red VW bug, without an inch of room to spare. They were going to New York, too, but can't take me. They gave me a number to reach them in NY. I got a ride across the border, and in Pennsylvania get a ride with a pilot, who has me drive his Volkswagen van on the turnpike to New Jersey, where I get another couple rides into New York City. I called my friends, then take the subway, which starts going above ground and delivers me to the Bronx.

The next day, I'm embarked on the Mauretania, a floating piece of Art Deco comfort. It has first class, cabin class and second class accommodations- first and cabin in the superstructure, and the second class rooms with portholes that look out from the hull. I have a four person room. I had booked with an acquaintance who was going to study as well, but he backed out at the last minute, leaving me the whole room. The first class area was off limits, but the rest of the ship was available to wander around in. Meals were on a schedule and seatings were with the same people each time. My fellow travelers were British, a bit older, and helped me out with the fine points of civilized dining.

A group of young people were going to France for study, and I became part of a singing quartet. There was a tenor from Baltimore with a characteristic accent, a soprano, and an alto from Bennington. I sang bass or baritone. We sang mostly from a book called the "Bach- Riemenschneider", a collection of cantata harmonizations, presented in score without words. It served as a treatise in harmony, with multiple workings of hymn tunes. We sang them "scat" style and as we progressed in our skill over those five days we began to acquire small audiences. The alto and I had a shipboard romance which did not survive our later meeting in Lyon.

I arrived and took a room in the hotel I had stayed at on my last night in Paris in '62, in Rue de l'Hirondelle. Between arriving in Paris, and the lessons I gave around the end of September, I needed to close a funding gap. I thought I might get work at the air base I had served at. I took a train to Evreux, and got a room near the station. The town looked different. A lot of rubble had been cleared and there was new construction. Bypassing the guard post, I went directly to the Canteen, where I knew the non-commissioned officers' club was. I talked to the Sergeant there and he was upset I hadn't checked in. As it turned out, hardly anyone was there.The base was being closed. De Gaulle was determined to close all the U.S. bases in France. There really was no work there. The sergeant, after conferring with his superiors, told me to simply walk off the base, the way I came, and not check in with the guard post. No complications.

I checked back into the hotel, and someone pointed out an office where American vets could get a small weekly stipend. Yogurt was recommended as the cheapest source of good nutrition, and bread was cheap in the boulangeries.

I wandered around town, seeing the sights, and drinking 40 centime espresso, 30 centime Beaujolais, and demi's of beer for 50 centimes, about ten cents. I gave a few lessons at the school and met some women who seemed to be connected with American interests who encouraged me in my musical efforts. It was still not time for the regular semester, and the canned sardines and sausages, and plain spaghetti at the cafeterias were not giving me the nutrition I needed. I made up a story about sickness in my family forcing me to go back home and quit the school job.

I went to the american Embassy and said I was running out of funds. Two military or agency-looking types told me if I contacted the Consulate in Antwerp, they could help. I began to hitch north. Near the airport I got a ride in a Citroen panel truck to Cambrai, near the Belgian border. Another ride to the outskirts of Antwerp, a picturesque town with small bridges arching over little canals. I walked through this district toward the center of town. I was hungry and saw what looked like a cafe with an open door, no sign, but the building decorated in psychedelic colors. I went in and saw a room with padded bench seats and long tables. No menu was posted. It was a kind of bordello. A woman came out of a door in the wall and I asked if I could get something to eat. She went back in, conferred, and it was decided that, indeed, it was time for lunch. People in varied costume, and equally varied sexual orientation, it seemed, filed into the room, and took seats. Food appeared and we all had a nice meal. My language skills were limited in French, and nonexistent in Flemish, so I didn't get much out of the conversation.

I continued on to the address I had been given, a two-story office building with steps down to an entrance into a partial ground floor. It was dusk, and the offices were not well lit. I was met by two men, again of military age and bearing, though not in uniform. No one else was there. They offered me coffee, which I accepted. As we talked the familiar feelings returned, and the grounds surrounding the offices revealed nature's ever present charm and comfort, if we would take the time to appreciate it. They said I could get a small ship, not a passenger vessel, leaving for America that evening, if I were willing to work as crew. I felt too sick and weak, probably from malnutrition, to take up the offer, and they gave me another address in Rotterdam. The main road between Antwerp and Rotterdam was full of traffic and it was not difficult to get rides.

Once in the city, it was a long walk out a finger of land going toward the sea. I found the place, a three story boarding house run by a Dutch family. Small rooms up narrow staircases with a day room upstairs and a dining room on a different floor. The meals were vegetables with butter soaked potatoes as the main course, and cheese. Hearty food which improved my state of health immediately. All the residents were Americans, of military age including some pilots.

One day, there was a big buzz. Someone with a lot of stories to tell was coming by on his way to the East bloc. He was going to hold forth in the day room up stairs. My security clearance was inactive, and since it was compartmentalised for nuclear medical info, I was not invited. I wondered who the mystery man was, and what were his stories. Did it have to do with Dallas?

A couple days later the decision was made to send me back by train to the Franco-Belgian border, from where I would "faire du auto-stop" back to Paris.

I boarded a train to Brussels, with an onward passage to Tourcoing. The countryside, instead of quaint old villages, consisted of large modern buildings set in the polluted remains of woodland that had suffered two world wars. Copies of a right-wing newspaper that criticised people who expected social welfare programs from their government were in a few bundles in the cars. It seemed to me they were fighting against something that had become mainstream in Europe.

I boarded my assigned car in Brussels after having been given a cup of coffee. There was only one other person on board, who got off after a short while. As I began to feel the effects, the car was shunted here and there, attached to different trains. I was used to being alone under the influence of one thing or another, and was bemused by the shuffle.

After the roundabout journey, I got off at Tourcoing, and walked to the border crossing there. The lights of the town late at night had a special glow, assisted by what was in the coffee in Brussels. It was not "A Starry Night," but it would do. The officer staffing the deserted building said I could sleep on the floor inside. When his relief came in the morning, he fast talked his replacement out of his misgivings and I crossed the border into France. I went into a streetside cafe where the proprietor served me a thick soup that made a host of troubles go away.

The textile towns in the French north had an air of returning to prosperity, with lots of consumer goods available, and a fairly optimistic attitude among the people.

Pigalle Follies

I got rides to the northern suburbs of Paris, where I was bothered by some right wing French youth, nothing serious, and where I took the Metro into town.

I checked back in at the Embassy. They said the two guys should not have sent me on the wild goose chase (the actual language they used). They arranged for me to be seen at the American Hospital, near the Bois de Boulogne. Staff there initially didn't want to treat me, but somebody overruled them. They then put me up in a nice building they owned near Pigalle. I was to rest and eat well.

I was recovering in a nice room, guest of the State Dept. I walked up to where the artists used to hang out, painters and wannabes set up with their easels and palettes in the open air cafes. There were still vestiges of the former rural nature of the place.

There were impressive vistas from the steps of Sacre-Coeur.

Around the corner from where I was staying, a cafe, where a jar of hard-boiled eggs sat on the counter. It's suggested that I try one that has an ink-blue tinge in one spot. The color of sandoz liquid LSD. I return to my room, lazily enjoying the sensations. From inside my armoire, a voice through a speaker said to go to a particular address. I wonder if I'm supposed to think I'm hearing "voices", but I know when I'm hearing a 3" speaker.

I went to the spot nearby, and found a brass plaque which named a violinmaker. I knocked on the door, and was admitted to a single room with no sign of violins, tools, or Gepetto. Instead, there were a bunch of clean-cut American types, including one older balding man whom all the others were deferring to. The acid had made me well aware of the pecking order in the room. The man, of a chunky physique, was having his ass kissed.

I took it he was Paris station chief or somebody else important from Washington. He took control and had me sit in a comfortable chair and get hooked up to a polygraph. No pentothal this time. This was my second and last such experience to date.

The questioning was brief. One question; “Did you do anything?” I took it to mean had I set up shop on my own, violated the Logan Act- legislation forbidding private citizens from conducting Foreign Policy, or taken upon myself to terminate somebody. I answered no, was unhooked and sent on my way with a paperback Ian Fleming novel, something where Bond finds himself in a missile launch tube and, of course, escapes.

I was hungry by this time, and bought a hot dog, a “Croque Monsieur” from a convenient street vendor,who insisted on slathering hot mustard on it. Dosed again… had a very nice walk.

Later, I received a ticket to Wiesbaden, Where I was met by an Army officer, who arranged a billet, and took me to a mess hall. When I got in line, I got a chilly reception, they didn’t want to serve me. There were grumbles among the G.I.s sitting at their tables. I knew what was awaiting them.

Vietnam was heating up, This was around the time of Tonkin Gulf, and they were going, and I was returning home. The officer gave them a lecture, saying that I had served honorably and for them to show some courtesy. After the meal, he and I went outside, to tree lined streets in the waning afternoon light. He produced a cigarette, which had apparently been doctored with hashish oil. I had been given something similar back at Lackland. We smoked it, bring a sense of well-being and peace. I remarked at the beauty of the place, I had expected Germany to have a bleak look, remembering footage of Berlin after the war. He shared how privileged he felt to serve there.

The next day I boarded a military train headed to Bremerhaven. We were cautioned not to be tempted to get off and stay in Germany as sanctions would be severe. As we went through an industrial area filled with warehouses, I thought about unwilling passengers twenty years before.

On board the Military Sea Transport Service ship, Our bunks, or racks, were 24" wide stacked four or five high along the walls, though the top row was not used. The latrine was in the bow, and as it was at the extreme, it moved up and down more than the midship area as the ship moved through the waves. The mess was on another level, midships. Meals were very good. Navy cooks had the best reputation of all the services.

Enlisted personnel were on three levels in the hull. The superstructure was reserved for officers, their families, and civilian DOD employees. Their quarters were spartan versions of a passenger ship's. In the hull, openings connected the three levels, so that equipment and freight could be lowered in. If one of these levels had been breached by a torpedo, hatches would seal off the damaged level, Those who could not be evacuated would face drowning, a fact that gave us a somber mood. Our daily routine included movies that would be shown in a large compartment, whose floor was the top of a large water tank called the cofferdam, on which we sat.

On the North Atlantic in winter, we were rarely allowed on deck. This was the province of the deckhands only.

I would play guitar in a stairwell that led to the better accommodations. A civilian engineer from topside would occasionally come by and listen. Another topside guest of the State Dept. was an American jazz musician who had spent three years in a French jail. He had entered the country with a little marijuana taped to his body. He made it through customs, but a few days later was arrested at his hotel room.

As it turned out, I owed money to the State Dept. for lodging and transportation, for which I received a monthly bill for the next couple years.

We were allowed on deck off the coast of Newfoundland and a cheer went up when those with transistor radios could pick up a country music station from the states. The terminal where we debarked was familiar to me from newsreels of G.I.s returning after V.E. day.

A trip to the Atlantic, and One to the Pacific

~

Returning home, life picked up more or less as it had been: teaching students basic guitar, playing classical at a few cafes, and time at the art museums- St. Louis' and Washington University's. I had philosophical and political discussions with other young people. I attended regular services at the Ethical Society of St. Louis, held in a deconsecrated synagogue near Grand Ave. midtown, called Grand Boulevard by the French. The Ethical Society had roots in New York, a sort of Jewish version of Unitarianism. Nominally atheist, but definitely spiritual and humanistic.

I was able to get counseling from the facilitator concerning questions of living an ethical life without the threat of eternal punishment. There was a local chapter of the American Society of Humanists, whose ideas my friends and I would discuss. Across the street from the Ethical Society was a store with products from Texas, including peyote buttons, displayed openly in the streetside window. None of my friends tried them, though we certainly discussed the phenomenon. I occasionally went to Mass at St. Louis Cathedral on Lindell Blvd. The cathedral was decorated in mosaics, a little unusual for a Catholic church, more like the Greek Orthodox institution in University city I also visited.

On the political side, I continued my studies, occasionally reading Muhammad Speaks, which I recognized as having little to do with Arab or Iranian Islam. One of our coterie was married to a Persian pilot who was in Iran at this time, and had a two year old child with him. She was a dancer, and sometimes performed to Persian recorded music at the local cultural society. I once accompanied her on guitar, to a mixed reception. To some conservatives it was too western and secular. Additionally she was with a man not her husband and not a relative.

At a party one evening we got an intel briefing, a conversation, about the situation in Iran. We supported the Shah, and there was an opposition. They were describing some of the horrible things going on. After a particularly grisly account, I said "wow, that's what we're up against!" and a guy said, " no, that's the good guys!" We were warned not to get involved, much less travel there.

Another time, another chat. We met some people at another party, saying the most critical, almost un-American things, stuff that would give Joseph McCarthy indigestion, had he not already gotten his comeuppance. It was pointed out that everything they had said was protected by the Constitution, which we had sworn to uphold. We were learning the position of what would be known as the non-communist left. I had asked, when I left active duty what I should be looking for, the influence of the Russians? I was told no, it was the East Germans. They had schools to train people to blend in as Americans. I met a Czech woman who seemed to be trying to bring me in. Once I parked near her place, near the airport, but before I went in, a guy called me over and said they were a cell they had been watching. I left and never renewed my acquaintance.

Late February 1965, in St. Louis, I got the news that Malcolm X had been killed in New York, which saddened me since I believed him to be much more reasonable than the members of the Nation. There were some mosques in the County affiliated with Malcolm. Near one of them, a white man had been killed. Knowing the pervasive casual racism practiced by some area whites, I surmised that he had just said things that seemed normal to him, but these "so-called negroes" were not taking it anymore.

Summer came and Sue Ann, and another woman, of partial Cherokee descent, and I undertook to visit that other woman's relative in Lake Worth Florida, near Palm Beach. I had obtained a Morris Minor, the size of a Volks, with a 900 cc engine. We got excellent gas mileage. We drove through Tennessee, tracing the route I had taken to Montgomery, into north Georgia, area of the old Cherokee Nation. We stopped at Ringgold in a scenic wooded area. It was Sunday and the good folks were having lunch in a nice lodge-like restaurant. We were tolerated for a time, but my beard, two women, one blond and one possibly Indian, proved more than they could take, given the peer pressure they faced. We finished quickly and retreated.

Arriving in Lake Worth, here was the Florida experience; sandy soil, rumors of alligators in the canals, houses with no interior walls- deliberate gaps to allow air circulation, pastel colors, and lots of flowers. If there was a point to our trip, I didn't see it. After a tune-up for the car we made a fast non-stop trip back.

Gaslight Square was changing a bit. It was getting more crowded, with people stopping there on their way to to "the coast," Los Angeles, and some to San Francisco. There were more experimenting with LSD, not the few who kept it quiet, but public advocates who linked it to "free love".

I was spending more time teaching in the suburbs, including the newly opened St. Louis County Jewish Community Center, and thinking about going to Berkeley, San Francisco seeming too high-pressure and expensive for me. Besides, only a bus ride across the bridge, Berkeley was where all the Segovia students were. A friend, and fellow musician, Gordon, had some source of LSD, and we had some discussions about it. I never got any.

Gordon had Hollywood aspirations, and I had a Ford four-door with good brakes and engine, and fair tires, so we went, in the spring of '66, with his girlfriend and another guy via Denver, Salt Lake, Vegas, and Barstow to the West coast. The Ford needed a couple tires and a generator by the time we got to Barstow. Since Gordon had paid for that, I called it even and let them have the car. I didn't really want the car expense in Berkeley. I started hitching north from Barstow and got a couple good rides that put me in Monterey, where I got an attractive offer from a woman to stay, but my mind was set on my original plan and I found myself on an onramp for Highway 101, toward Berkeley.

It was March when I arrived. I stayed with a woman from our group in St. Louis. I got the lay of the land, found accommodations in a building with separate rooms, shared bathrooms and two kitchens. I met lots of new faces, some with interests in common. The building was full of people exploring the then new world that Tim Leary and others were touting.

Across the street,there was a convenient music store with a good supply of students, so I started making enough to live on. I found a campus YMCA, Stiles Hall, where I began teaching classes in folk guitar, and where interesting authors gave lectures. I listened to Eric Berne, author of *Games People Play* explain his idea of transactional analysis.

I became acquainted with Telegraph Avenue, with its bookstores, Cody's and Moe's, a store with imports from India, and a Persian restaurant, Reza's. Years later, I went in for a meal one day and found it full of Arab looking men with four inch beards. They said the place was closed. I questioned them and asked where the usual employees were. Out of the kitchen came a tall man who looked like the man we would later know as bin Laden. They tried to explain their new attitude about America. I tried to say to them that, yes, we may currently have a right wing government (Nixon), but that we were working to change the direction of the country. This brought them no comfort, and they said it didn't matter who governed us, America was corrupt and must be destroyed. I guess they felt no threat from the hippie who questioned them. I left the place, and wanted to blow the whistle, but by this time had no connections to anybody in security. I informed a Berkeley policeman, but doubt anything came of it. Some years after, the Iranian takeover of the American Embassy took place.

The Mediterraneum and the Forum cafes were sites of late night discussions, fueled by espresso, cappuccino, and caffelatte. Telegraph ended at Sproul Plaza, where the Free Speech Movement had its genesis. I found my way around The University of California Campus with its creeks and groves, a green oasis in the little city.

The counterculture (the term was not then in wide circulation) newspaper, The Berkeley Barb, and The San Francisco Oracle provided the latest hippie news, and information on the nascent anti-war movement.

Over the bridge, San Francisco, with Haight Street; the "panhandle," a grassy strip nearby, where bands would play outdoors; and Golden Gate Park were sites of pilgrimage. North Beach was home to Italian family-style restaurants and City Lights Bookstore, focus of censorship controversies. Herb Caen was the columnist who reported all of this to daily readers of the San Francisco Chronicle. Mention in his column meant you were somebody. One particular somebody was Bill Graham, the impresario behind the Fillmore Auditorium, where the Grateful Dead played to a light show. Graham promoted the bands that became the soundtrack of our times. The Dead were playing one such concert, and after one set, most of the players left the stage, leaving two of their number jamming and keeping up the flow, a typical G.D. technique. The rumor on the floor was that they were backstage "doing acid". The whole band began playing again, and as the effects kicked in there was a palpable change in the music and we were in for a wild ride.

Berkeley drew affluent teens and young adults from the suburbs of Canyon, Orinda, Lafayette, and Concord, who became the first "hippies." They were joined by a wave of youngsters from all over the country who would redefine the term. Near the Berkeley campus, I met a wiry guy talking nonsense from whom I distanced myself. I later recognized him as Charles Manson.

One day in the kitchen of my building I would meet a slightly older man and a fourteen year old hanger-on. He was known as Augustus Owsley Stanley III . He was cooking some meat in a frying pan; no sauce, no oil. He believed in eating only meat- a pioneer of the "paleo diet." He lived over the hill in Orinda, a town with a regulation that if you cut down a tree, you had to plant six more. Owsley was a chemist who could synthesize LSD. At least, this was his reputation. It is possible he could have had another source. He marketed doses of 1000 micrograms, at the time in a powder form in tiny capsules, "Owsley acid."

One evening, about a year before the "summer of love", nearly everyone in the building took Owsley's capsules. We enjoyed a night of music and animated conversation. "Deja vu" experiences were discussed. I don't remember anyone "freaking out", probably because of the social solidarity of common experience. Nobody felt isolated.

As I mentioned, during the late fifties, an Army major ran experiments in S.F., where prostitutes' "johns" would be given doses in drinks and the proceedings would be observed through two-way mirrors by researchers. Aerosols of LSD were also released outside the Golden Gate to drift in with the fog, though I doubt this could have any measurable effect. These doings are documented in, among other publications, Acid Dreams by Martin A. Lee, who mentioned an old OSS man who facilitated LSD's distribution, and who I met briefly. Later there was a certain amount of open drug dealing in my building. A few weeks on, a DEA raid netted these newcomers. Not one word was said to me.

LSD became illegal in 1967, and years later Owsley was a guest of "Club Fed", along with a sidekick I knew as a musician.

That year, one of the three leading lights of the dubious Federal Project MKULTRA, along with Allen Dulles and Dr. Sidney Gottlieb, Dr. Donald Ewen Cameron, whose repulsive social ideas led him to torture mental patients in a Montreal hospital under the auspices of our CIA met his demise. Wikipedia has a good summary of this worthy gentleman's wake of destruction.

In Berkeley, down Telegraph Ave. from the University's Sather Gate was a club called the "Jabberwock." The house band was Country Joe and the Fish, Joe McDonald and Barry Melton the frontmen. Country Joe was a Navy vet, having served in Japan. He later became a veterans' advocate. The band's signature song was "Fixin' to Die Rag"- "Come on all of you big, strong men, Uncle Sam needs a helpin' hand- he's got himself in a great big jam, way down yonder in Vietnam…", to the tune of "Muscat Ramble." A jug band-style tune that became ubiquitous during the protest movement. I started noticing how many vets were showing up for the demonstrations, and I began taking their side.

At the club I met some people who were starting the Clear Light Society, named after a Buddhist concept from the "Tibetan Book of the Dead." I heard Richard Alpert speak, one of the original researchers along with Tim Leary at Harvard. Alpert was just back from India, where he studied with a Hindu guru, and was backing away from chemical means to transcendence.

I later attended a house lecture near the stadium given by Leary. Interestingly, he had done a couple of years at West Point, but was not disciplined enough for the Army. He may have retained some connections from those days, but that did not save him from a stint in Club Fed for a while; he escaped and ended up in Algeria as a problematic guest of the Black Panthers. This one of his lectures was his standard "Turn on, tune in, drop out" message, but with a futuristic look at how computers might be a part of the effort. Anyone seeing people walking around today with their heads down, staring at their iPhone screens will see that Leary was prescient, though maybe not in the way he thought.

September '66, I begin a semester at Merritt College on Grove St. A class in ethics and one in film history stay in my memory. I was apparently still under my Air Force orders since two guys came up to me between classes, military types, and gave me an opened bag of marshmallows and told me to share them with my classmates. They had the telltale blue where they have been injected with liquid Delysid. I had them on my desk, and a classmate asked me for some. I tried to discourage him, telling him they are "special." I persisted in my warnings but finally give him one or two.

Not much later he had apparently freaked out, was taken to the hospital. Two people from the college come and ask me about it. I tell them truthfully how I got them and how I warned the victim. They confiscated the dubious confection, but may have been given a heads up by the military, so I escape any consequences. Remembering back to an earlier incident at my building; Some fellow there had already taken some and had a mixed reaction. He was offered some more days later and took it. He began to get second thoughts about it and swallowed Drano, an alkali, to counteract the acid. The presumed neutralisation did not occur and he also was hospitalized, still high. Two things distinguished these unauthorized experiments from those I had undergone in the service: one was the very large difference in dosage. There is a saying "the dosage makes the poison;" 85 vs. 1000 mcg. The other was the structured environment in the military, and the supervision by mostly well-meaning people.

I had some good teachers, but dropped out before getting credit- darn you, Tim Leary. At Merritt, I met my future wife, Marty, and I move into her apartment in the Rockridge section of Oakland, now a mostly unattainable upscale area. Marty is disenchanted by the drug use she saw around her. Under her influence I steer away from psychedelics and pot.

We are hikers and nature lovers, no vices but cigarettes and strong coffee. Her grandfather was an Army General, then retired with a large ranch bordering Hwy.101 in San Jose. Her uncle was an Air Force major, and family dinners reflected the division that Vietnam had caused, both of us seeing folly, him seeing duty.

In 1967 we move to Maui, with the intent of staying. We lived in Lahaina, in a building owned by Sam Farr, the Monterey congressman. He had a watering hole, the Farr Outt Bar on the first floor. I got a playing job at the Sheraton, Kaanapali Beach. One night after work someone bought me a

Black Russian, Kathryn's favorite drink. Hitching back to Lahaina, I started feeling the old familiar. Somebody has gotten to me again.

We had shipped over bikes and a Honda 90 motorcycle. Rides around the island and meals by the water made for an easy tropical life.

One day a guy told us about someone in Honolulu who claimed to be John Lennon, although he looked nothing like him. This merely seemed strange. Years later an eerie coincidence.Mark David Chapman, Lennon's assassin, had been in Hawaii at the time.

Paradise eventually became constricting and we relocated to Berkeley and all its social ferment.

We had heard Martin Luther King speak on campus in 1967. On April 4th, my father's birthday, in 1968, we heard the news of his death. That and the My Lai massacre had American youth in full rebellion. France's young people, too were in the streets that year, although the Gaullist Party won the election handily.

In May of 1969, I received a letter from my draft board informing me I was now classified 5-A, military obligation fulfilled. I was now free of all the military orders I had been given. I was still under the forty-year classified information secrecy obligation, but I was beyond further military orders.

With the sense of freedom I now had, we started making plans for me to study classical guitar at the "Staatsakademie fur Muzik und Darstellende Kunst in Wien," and for her to study dance at the school at Schloss Schonbrunn there. I applied and would be admitted if I passed an audition at the school. I filed for the V.A. educational benefit- I was eligible for 11 months of payments, and began saving for the trip. We would go as tourists initially, and apply for a long term visa if I passed.

We had booked a flight on Icelandic airways, the budget carrier of the time, to Luxembourg, and would go by train from there to Vienna. We got to Montreal, but Icelandic didn't have a suitable plane for the next leg, so we were rebooked on SAS to Copenhagen. Helpful trilingual Stewards served the passengers' needs. A stop at Keflavik debarked passengers headed to Iceland, and those making connections there to Shannon, Ireland. As we approached Copenhagen, fog made a landing impossible, and we were redirected to Hamburg. A long train trip ensued, changing to an Austrian train at the border at some remove from Salzburg. At one point in Germany we had shared a compartment with a young G.I. who was imbued with the necessary sense of the mission in Southeast Asia. Awkward conversation followed. Marty and I were, by that time completely disenchanted by our government's strategy, or lack thereof. When the young soldier left, we shared a sense of foreboding at what might be his fate.

The train on the Austrian side was pulled by a steam engine. It was explained that it was best suited to the mountain terrain.

After arrival in Vienna we found a hostel on the north side of town, where vineyards shaded into housing. Outdoor cafes would welcome diners to sample the recent harvest- the "heuriger," the wine of the day, sometimes with stems and bits of leaves from the hurried first pressing

We were befriended by an American couple, a photographer of Polish descent and his wife. They had recently arrived from Czechoslovakia, after the "Prague Spring" had been derailed by the Soviets.

Zeni had made a living in the U.S. as a professional photographer, following golf tours and providing flattering pictures of golfers in action. He had also been in Poland documenting ordinary people and their daily life. He had a Hasselblad camera with a waist-level finder- this meant you looked down into the top of the apparatus and saw the image on ground glass and moved the camera to compose your shot. He showed me how to turn the camera sideways and appear to be fiddling with the controls as you got a candid shot to your right or left. Zeni and his wife were in transit back to the U.S. We later caught up with them in Santa Barbara.

Our first order of business was to find housing. We tried various agencies with no success. Somehow, Frau Erika Hofer came to our rescue. She had been an analysand and student of Sigmund

Freud. She had an apartment in the central area that may have been his. She showed us his couch. It was covered by a couple of oriental carpets.

She was the head of "Die Internazionale Frauenliga fur Frieden und Freiheit," the International women's League for Peace and Freedom. We had the Peace and freedom party at home. More importantly for us, she had a house in Neuwaldegg-"the new woodsy corner" in the northwestern suburbs, where she lived with her octogenarian husband, also a psychiatrist, when she wasn't practicing downtown. They thought it was too noisy there. The house next door was said to belong to Ursula Andress, but I never saw her there.

Neuwaldeggerstrasse was just past the last stop on the Hernalser Strassenbahn, the line that ran along Hernalser Hauptstrasse- "the main street of the Lord of the Als" - the small river that paralleled the street. It was a pretty prosperous shopping district. The line passed through some working class neighborhoods and ended at the Ringstrasse,which had been a defensive city wall in the 1600s and now was a wide boulevard on which were located the city's most important landmarks; the Rathaus; the parliament; Museums; parks; and the Opera, the crown Jewel of the city's cultural life. A streetcar on the Ringstrasse would bring you to to the Akademie's main building.

West from Frau Hofer's house extended the Wienerwald, the Vienna woods. The lands had been the property of the Hapsburgs and other nobility, but now belonged to the city, die Gemeinde. Especially on weekends, people would stroll the many paths through the former palace lands and hunting grounds. If you lived on its edges you could take roundabout routes to other parts of the city. We were privileged to live among the mix of old upper middle class homes and city owned low rise apartments. In one direction were the spooky ruins of an old palace. It looked like it had been bombed out, but neighbors told us it had been the property of a particularly cruel Nazi official and, in the last days of the Third Reich, people had stormed the place and reduced it to ruins. The past was not too long past in parts of the city.

In another part of the woods, stood the Jubilaeums Tor, a small tower monumentalizing a war before the Great War. It looked a bit forlorn, but was a point of local pride. To the southwest, there was a ski area with lifts, closed in the autumn, in a nice area with pine trees. Most of the Wienerwald was deciduous. Nearby was a meeting area with tables and benches in rows facing a stage or dais. We were told that right wingers still used it for speeches and rallies, and warned not to be near when it was occupied.

In another site was a military cemetery, a somber place where we were reminded to avoid levity. It had fallen soldiers from the two big wars. It was a bit like Japan's Yasukuni Shrine.

On a lighter note, near us was a neighborhood of weekend houses with small orchards where people would place baskets of their produce on their front steps, paid for on the honor system. The American High School, where I gave a few guitar lessons, was in a clearing a half mile walk from Frau Hofer's.

Similar weekend garden sheds were dotted through the woods, some with meadows where animal fodder was grown, bearing signs saying "futterwiese, nicht tretten."

As was usual in many parts of the German speaking countries, shops were open nine to five, until noon on Saturday and closed on Sunday. If you weren't careful your pantry was empty on the weekend. Daily life was school, shopping, Strassenbahn, and strolling in the woods.

Marty went to Schoenbrunn, I went to the academy, passed our interviews and settled into a routine. I was authorised for the music education program, but the professor who auditioned me wanted me for his Konzertfach or concert specialty program. This was a pretty intense class- three days a week, eight or nine hours a day of Master Class. Prof. Karl Scheit's twenty-seven students had to attend all three classes, each of us getting an hour session a week.

During the lessons the professor would comment to the student on points of technique and interpretation. Each of the rest of us was expected to absorb the lesson and not make the same mistakes when it was our turn. Each student built up a repertoire of pieces that evolved through time. On the other days, we had German language and culture lessons with the other international students. A Japanese woman, in the spirit of "cosplay," once wore a Kimono to class, and on another occasion wore a "dirndl." Lederhosen and "tragenanzug", a kind of hunting uniform for outdoor activities were available at a Department store downtown, but I was not tempted.

Guitar students were expected to practice five hours a day- not too bad- violinists were in for up to eight.

In addition to their devotion to the early wine harvest, the Viennese valued the coffeehouse. The story is told that after the siege of the city by the Turks, the Ottomans in their retreat, abandoned their stocks of coffee, and the equipment to roast and brew it. The tradition in Vienna is brewed coffee rather than espresso, usually with steamed milk, like a cappuccino, and always served with a glass of water. Racks have the daily newspapers and you are not pressured to leave, even if you haven't ordered another cup. People did not abuse this however, and while laptops and cellphones did not then exist, you could linger over a book, or a newspaper.

We were invited to a meeting of the Frauenliga. The State Department, on our arrival, had told us to avoid any anti-war protests, and we had the point reiterated at a meal to which we had been summoned by some important looking government type. This however, was Frau Hofer, and not a demonstration. We met some young people there, mostly women, and while we considered ourselves typical Berkeley liberals, here we were exposed to the full force of European anti-Americanism. This was tempered by an awareness of the U.S. as a counterweight to the Soviets, who found absolutely no support from them.

Here and there through the woods were small restaurants serving hikers pastries and soups. On one of our walks, we saw what looked like one, with a sandwich board sign announcing the day's menu. We tried to get served but got the cold shoulder. We had been welcomed or at least tolerated in every other one we had visited. The people did not look Viennese, nor any other western European nationality. I had the feeling they were Russian, and in fact a Soviet office was not very far away. They could have been taking a break from their cold war duties. We didn't stick around to find out.

Throughout the woods were craters from Allied bombs, mostly on an axis paralleling the rail lines that came in from the west, the northern sections being too hilly for practical train routes. In these craters, water collected and by summer's end roses bloomed at the bottom, water loving plants that they are.

Once, we were following an unfamiliar trail. We saw a clearing, and a gravel track leading to an Institutional building and a driveway leading to a parking lot behind it. We started down the path leading to a curb, and suddenly a security type came up to us in a rush. He said a VIP was coming. We came to the curb as a bunch of well dressed people were lined up there with their security personnel. Our guy explained that the head of the Palestinians was arriving. Dumb as a post, I said "David Ben Gurion?" We stand at attention, us in our hiking clothes, as Yasser Arafat steps out of the back seat of a limousine, and looks right at the two scruffy Americans with a "who are these mooks" look on his face.

Quickly, he is hustled inside. Had our government's policy been different, someone would have had a perfect line to a shot as he exited the car. It was not his day. Fortunately for all concerned, a law had been passed in 1967 forbidding Agencies, Bureaus and Departments of our Federal Govt. from killing foreign leaders.

On the Road, Again

~

A transition occurred in my interests. Before the service, and after I got back to St. Louis, my focus was about equally on music, art and photography. When I moved out to California, I left my paints and camera equipment at my parents' house. My concentration was almost completely dedicated to my musical studies. This continued for many years. I took up painting again in the eighties, and only revisited photography a few years ago, after the advent of digital cameras, and the fall in prices of all the film cameras I couldn't afford in the sixties.

Studying music history more carefully I began to see all history measured or paced by music. Gregorian chant in the middle ages, The Notre Dame school around 1200 with the "cantus firmus", a basic melody elaborated by quicker notes in the upper voices and the flowing counterpoint of the church composers from Josquin de Pres to Palestrina, 1400 to 1500,Elizabethan lute music, the Spanish music that accompanied the conquistadores to the New World, The Baroque and early Classical music that inspired the enlightenment thinkers who produced our Constitution and the French Baroque styles that led New Orleans to develop Jazz.

Closer to our time, ragtime before the Great War, swing in the thirties and forties and the big band sound of WWII. The jazz combos of the postwar period and the anti-war soundtrack of sixties rock. Disco after the fall of Saigon.

The links between the military and music go back a long time. The cadence of marching troops sets the timing for the order of battle, punctuated by drum and bugle signals for battlefield communications. At Colonial Williamsburg, the music teacher's house was occupied by a former military man, a man who could keep time. Musical rhythm was formed by marching and dancing.

Winter break came and we decided to go to Paris to meet my friends Michael and his wife Jonnye Faye. Marty and I had done a drive-away trip back in 1968. You transport someone's car in exchange for the use of it for the trip. You don't get paid, and you have to pay for gas and leave a deposit, as well as getting a doctor's certificate. Still, a step up from hitchhiking. We left from San Francisco and would drop a '67 Chevy pickup, in Tulsa, then hitch the rest of the way to St.Louis.

The truck got surprisingly bad mileage on the way- 11 mpg. Our coffers were seriously depleted. Upon delivering the vehicle and handing it over at a hotel in Tulsa, the owner tried to get away with not paying back the deposit. I was not going to let this white Oklahoman exploit yet another Indian. Figuring "qui tacit consentit," I made a loud stink about it in the lobby, and apparently swayed the tide of of public opinion, and he coughed it up. We then hitched the rest of the way into Missouri, where a driver turned over the wheel of a rear-drive Corvair to me.My earlier experience driving in snow on bald tires served us well. With good tires, and the weight of the engine on the back wheels, I kept us on the road and delivered us to St.Louis.

This experience hitching in winter added to the other trips I had made during that season gave us the confidence to try it to France. We made it through Austria and Germany to northeastern France, where it was warmer and the roads were clear. We got a ride from an American officer on some of the last of the traditional tree-lined highways. They had a hazardous reputation and were being phased out. We came to a small city on the way which still hadn't been rebuilt, even in 1969. The outskirts were still in ruins. The center of town, however had been reconstructed, and soon we made our way to Paris.

We met Mike and Jonnye. He was studying at the Schola Cantorum, taking lessons from Alexandre Lagoya, who along with his duo partner, Ida Presti were considered the best in Paris. Jonnye was working at UNESCO. Michael's guitar style was excellent, full of varying tone color and I considered him the better player at this stage of our rivalry. I would double down my efforts in the coming semester.

They lived in a building on the right bank, right on the river, with a million dollar view from their balcony on the third floor, just upstream from the Mairie, the city hall. The district was called the Marais, the swamp, but at the time it was among the best neighborhoods. The building was called the "Cite des Arts" and was restricted to a few people selected by a board of directors. Among the residents was a relative of the Shah of Iran, who contributed large sums, probably of our taxpayer dollars, to the Foundation.

We visited Jonnye's office at UNESCO and its gardens. After a week or so, we returned to Vienna, hitchhiking in better weather than our outward trip. Our ride into the city from Sankt Polten, a suburb west of the Wienerwald was with an older gentlemen who warned us of the dangers of hitchhiking in the area given the extra dimension of cold war tensions added to the usual element of criminality, and pointed out some landmarks of the German occupation. He had the air of a retired Austrian intelligence officer.

After a productive semester we returned to the States, again on Icelandic, then by bus to Berkeley. At Cleveland, changing buses, I had another suitcase stolen, but the things I had of value from that trip were safe.

I found work at the San Francisco Conservatory of Music, at 19th and Ortega, where I substituted for George Sakellariou, who took a two-year sabbatical. At the time, commuting from Berkeley in the afternoon was not at all a burden, compared to today. I was nominally in the "Prep Department" which meant I taught kids after their public school day, although I had as many adult students. I gave a classical guitar recital there. The atmosphere in the Sunset District was pleasant if a little foggy. Across 19th Avenue was a small grocery store where students and teachers could get food and coffee. As one sat on the steps in front of the school, the window of the store seemed to advertise "ZEN FOODS". The edge of the entrance blocked the view of the "FRO" at the beginning of the sign.

It had been quite a decade for me, and for our country.

Appendix l, A Plausible Bit of Fiction

~

A CIA Director, Two Doctors and a Colonel Walk into a Bar….

It is later in Eisenhower's second term. Allen Dulles, who negotiated the surrender of Axis Forces in Italy, earning the gratitude of many in and out of government, sits in a cafeteria near Boston Public Garden frequented by area professionals and academics. He is Director, Central Intelligence Agency. Dulles thanks Dr. Ewen Cameron for coming down from Montreal. Sharing the table are Dr. Sidney Gottlieb and an unidentified Air Force Colonel.

"We need a lot more data", Dulles said. Cameron said "Our research is going well up in Montreal, I'll write a summary of our work and get it to you. Between the driving" ("psychic driving,"where the subject listens to snippets on tape in conditions of sensory deprivation"and the dosing...", "Delysid", Gottlieb interrupted, "Dr. Hoffman's discovery during the war, in Switzerland, also called LSD, is very promising for interrogations….". Sidney pulled out a small bottle filled with a blue liquid with the Sandoz brand. "Well", said Allen, "your subjects are all mental patients, and we need more information on how it might affect normal people." "I have a fellow at Harvard looking into the religious or spiritual effects of it. Apparently some get that out of it. "He checks out security wise, a West Point dropout, spent the war as enlisted- a psychologist a Sergeant! The Academy thought he was too wild, but for us… his colleagues have no idea. Sandoz has given us exclusive access for now."

The colonel piped up; "We have a single basic training site- Lackland in Texas. We could get you a lot of good subjects- all volunteers, no draftees, they test higher than the other services, but we would need a parallel chain of command- no sense in spreading it around- I know some officers would object. Sir, what exactly are we trying to prove?"

Director Dulles replied, "Several things. Will it enhance or degrade performance?, is it good for interrogation? Can we use it for our people to resist?" "One more thing- can we use it to produce a "super soldier"- maybe training in combination with hypnosis so we could trigger an action that the subject would retain no memory of, at least that our adversaries could get out of them?". This idea is from Advanced Research Projects at DOD and our own Technical Services"

The Colonel asked; "does Ike know?"

"What he needs to" A.D. replied. "The money's no problem. Let's meet sometime next year, when we have more to go on."

The Devil's Dictionary (With Apologies to Ambrose Bierce)

For most of us baby boomers, (and war babies), knowledge of the common so-called psychoactive substances is commonplace. For today's hipsters a new crop of designer drugs is their common knowledge, of which I am blissfully unaware.

In this narrative I have mentioned, sometimes without background, various drugs. Here I offer an ABCs of some that were current during our young adulthood. To wit:

Alcohol; depending on dosage, a social and conversational stimulant. At higher levels, an anaesthetic and disinhibitor. Generally available to teens in parents' liquor cabinets, generally diluted with water after unauthorised use to maintain level in the bottle. Considerable potential for abuse. Beatniks and bohemians preferred red wine, a very good red from Israel, Binyamina, available in St. Louis liquor stores.

Amphetamine; a stimulant, popular with Air Force pilots, students studying late and musicians. Dexedrine the preferred brand, potential for dependency. Superseded by crystal methedrine, a Nazi drug still causing trouble.

Aspirin; an experimental drug for younger teens, mixed with Coca-Cola; ineffective. No abuse potential. Myth started because of similarly named drug from the twenties, aspirin, phenacetin and cocaine.

Benzedrine; a combination of a barbiturate and amphetamine, reputedly popular among jazz musicians of the be-bop era. May be reflected in the music. Repeated use may lead to seizures, may have been part of Elvis' armamentarium. A related compound, sodium pentothal, now called sodium thiopental, (if it were advertised today by Big Pharma might be called Thiopental Sodium) is handy for interrogations and executions. In fact, administered intravenously, could make the subject feel he is dying.That may enhance its use as the "truth serum."

BZ; the U.S. Army's name for a drug related to nerve gas, administered in sub-lethal doses. An incapacitant, developed as a cluster munition, for use on the battlefield, now prohibited by international agreement. Nerve gas, familiar to veterans of the great war, is a related compound to insecticide. A factory that can make bug-killer can make nerve gas.

Cocaine;virtually unknown in my circles. Otherwise, needs no introduction.

Codeine; a weaker form of opiate, subject of a good song. Buffy Ste. Marie has a version.

Dextromethorphin; over the counter synthetic opiate, in cough medicine. Cheap high with abuse potential.

Ephedrine; over the counter antihistamine, initially derived from a southwestern desert plant-Texas tea. Pleasant as a tea. Popular in Utah. Used by tweakers to make meth.

Heroin; well known in the U.S. as something to stay away from, unaccountably popular among some British rockers. Screw you, William Burroughs.

LSD; as stated at the beginning of this volume, a class of compounds derived from the ergot fungus, a blight on the rye crop. Medieval times saw regular outbreaks of St.Anthony's Fire among the peasants who could not afford bread made from wheat. While ergotism includes many dire physical symptoms, LSD's effects are mostly mental. The drug is effective in small doses. 10 mcg. is effective for depression and has been studied for treatment of alcoholism with a daily dose.The dose in the Air Force program I participated in was 85 mcg. The dosage packaged by Owsley Stanley in the mid sixties in the SF Bay Area was 1000 mcg. Ken Kesey's "Acid Test" had the drug mixed in Kool-Aid, which yielded wildly variable doses. Lysergic Acid Diethylamide-25 was the chemical name and Sandoz Pharmaceuticals branded it "Delysid.

Marijuana, marihuana, MaryJane, pot;- all were the female flowers of the hemp plant, which for centuries provided rope for sailors and farmers until 1937, when dacron producers prevailed on the federal government to ban it to help establish their new industry. This was the era of Reefer Madness and other propaganda. In Berkeley, mid sixties, the usual pot was the female plant, culled from field grown agricultural hemp, usually from Mexico. A pound could be had for about $25, making for a good profit with ounces going for $10. Pot from Colombia, Panama and Jamaica was rare and prized. The Berkeley Barb newspaper printed the weekly wholesale price.

Morning Glory seeds; oliloqui, known to the Indigenous Maya in ancient times, only the white and blue varieties could be used. The seeds had to be crushed for absorption.

Nitrous Oxide; familiar to dental patients. Acts by replacing oxygen in the brain, so is mixed with oxygen in dental practice. Known as laughing gas. Available to hippies at the time in whipped cream containers. Today's whipped cream cans use an inert gas.

PCP; a hog tranquilizer for slaughter. Had some popularity, led to unpredictable behavior. Could partly explain the Manson Family's actions. Sometimes used to "enhance" weak marijuana. Use highly unethical by both seller and user. Responsible for at least one suicide death on the Berkeley campus in recent memory.

This partial list is not intended as a how-to, and does not imply advocacy by the author.

Notes on the Chapters

~

Military Roots; The world in 1943: Our country is united in a great effort to defeat fascism. Our enemies are Germany, Italy, and Japan. Before the Pearl Harbor attack by the Japanese Navy, It was not at all clear that we would oppose Hitler. A large number of people here did not oppose the idea of "white" supremacy, plenty hated Jews and communists. Many thought that union members were communists and many hated Franklin Roosevelt, and his "socialistic" policies to help the U.S. out of the trouble capitalist excess provoked.

By '43, there was beginning to be a consensus which brought together our disparate ethnic groups, businessmen, union and non-union workers into a common wartime culture with a big band swing music soundtrack. Women were working as equals, more or less, in the defense industries. Discrimination existed to be sure; if you wanted to be a pilot, you would have to change your Jewish sounding name, and African- American soldiers were led by white officers.

The country was still agricultural as well as industrial. The Army brought together farm boys and city slickers.

The nation was spanned by extensive and efficient rail transport for both freight and passengers. We were further linked by radio and newspapers, and our rotary dialed telephones enabled anyone to call anywhere in the country. For text messaging we had the telegraph system. Military and civilian aviation developed simultaneously.

This was the country I was born into. 1943 saw scientists at Los Alamos on the brink of perfecting the weapon which would, at the end of the war, announce to the world a new threat that would be at the core of the Cold War, which would turn allies into adversaries.

1961 brought a new President, a former Navy lieutenant to replace the general who had presided over the transition from W.W.ll to the Cold War.

We had suffered through the time of the House Un-American Activities committee, and the fall of its chairman, Joseph McCarthy, and his slimy Lawyer, Roy Cohn.

A new generation of optimistic, mostly tolerant American men and women were ready to play their roles in our military services.

The Green Monster; A one story wooden building extending over a large area across the highway from the Basic training barracks', painted green, was said to swallow civilians at one end and spit out airmen at the other. Uniforms were issued and testing conducted there.

Road Guards Out! I had not heard of two things: LSD, and the saying among servicemen; never volunteer.

Chain of Command; Actually, above the resident is another level of command. The very top of the pyramid of power is occupied by "We the People of the United States of America". That's a lot of people for such a small area. (See: How many angels can dance on the head of a pin?) The power of the people is exercised through the Constitution by the elected representatives of the people, the Congress, and occasionally directly by the people. If the Congress declares war, the president, as head of the executive branch, is charged with carrying out actions in furtherance of congress' wishes. Many modern presidents gloss this over. If there is a conflict between congress and the president, or between the public and either or both of these branches, or between a State or local government, the Judicial branch has the remit of sorting it out. The courts tend not to wish to insert themselves into military matters.

There are problems that arise in this system. One is that the executive may look the other way and not exercise appropriate supervision, and relatedly someone down the line will substitute his or her own judgment. A luminary of the stature of Kermit Roosevelt, in charge in 1953 in Teheran, did exactly that by ignoring State Department instructions to abort the planned coup against

President Mossadegh. Perhaps we could have spared ourselves the Iran hostage crisis.

The Rifle; this weapon plays a central role in the Army and the Marines. Airmen, by custom are only expected to defend their bases for twenty-four hours, before those other two services take over. Rifle training is therefore less extensive than those services.

The Fraternity; I use this term to indicate someone answerable to some degree to some intelligence institution. I don't mean those professionals who walk in the front door of those agencies, departments, bureaus or offices. In my experience someone on the bottom rung will not know who exactly they are answering to. I never checked anyone's ID. If they knew to tell me something I assumed they were authorised.

Not With a Bang; Every time I was dosed during basic training, I took pleasure in mastering whatever distress or discomfort was present and doing to the best of my my ability whatever I was tasked with. I was definitely looking forward to our graduation march and was very disappointed at its cancellation.

Medics; The sense of personal freedom was the biggest contrast between basic and basic med. The preventive medicine course was a rewarding intellectual challenge, and I have treasured the knowledge and orientation I gained from it for my entire life.

Chemical, Biological and Nuclear; This study section impressed me with the seriousness of the effort to avoid nuclear war in the first place. I could see that nuclear weapons posed the highest order of danger to public health. Taking an intelligence test with a fairly stiff dose of LSD was, in retrospect one the highlights of my life since my score reflected that I hadn't screwed it up.

La Belle France- the Good and the Bad; The chapter perhaps emphasised the bad a bit. Looking back, learning another culture at a small town level, and having access to the "City of Light" was a great privilege, as was knowing the young Americans and British I met off base.

If You Can Make it Here; My father was a child of Jersey City. The rhythms of the speech and attitudes were already familiar to me, but I was intimidated by the scale of the city. My decision later to settle in Berkeley rather than San Francisco was influenced by my first exposure to New York.

Summer of '62; Echoes of France in St. Louis were more apparent to me than before and I found plenty of cultural opportunities there. My friend Charlie was done with his active duty in the Army, and we kept busy.

Charlie and Mike's Excellent Adventure; The story about Corning, Ark. from our high school years that may shed some light on the paths we followed later.

The Midnight Ride of Michael Paul Revere; The requirements of secrecy and my instructions to forget my experiences left me with a kind of restlessness, and wanderlust came on me fairly regularly. My experiences in New England gave me a wider perspective, and a finer ethical sense. The rockbound Maine coast was beautiful, and Ogunquit jail was another useful rite of passage.

Music and Mexico; I met more veterans, Michael was Navy. There were some Air Force vets and women who may have been vets or otherwise connected to the military who became my circle of friends around Gaslight Square. The trip to Mexico was a great group experience and added yet another perspective.

Gaslight Square and That Day in November; Like the way everyone remembers where they were on September 11, 2001, John F. Kennedy's death on November 22 was the watershed event in our lives. A lack of the hope he had inspired in us dragged us down.

Looking in on the rise of the Nation of Islam was another perspective.

Back to Paris; Why did I consent to go with the possible Klan members to the Indiana State Park? Was I simply not assertive enough? If I had refused, along that then deserted stretch of Highway 40, they could have simply forced me to go with them. Was there something in their manner that lessened my concern? Why did my friends from St. Louis come along exactly when they did? I was actually used to not asking too many questions, when at any moment my perceptions might undergo a shift.

My time aboard Cunard's Mauretania; five days of music, moonlight and magic: Brigadoon in mid Atlantic.

I was surprised to find the air base where I served nearly deserted. I had hoped to wander around and recover memories of my interrogation at the "bakery." Walking around the town brought little to jog my memory since the town had changed a lot in the two or three years I had been gone. Paris was still itself, but my lack of financial planning squeezed my enjoyment of it.

Air travel gets you from one place to another. Trains offer a sealed in view of the passing landscape. Driving gives a better look, but hitchhiking, with the time spent at the side of the road, and the walking through towns to find the best place to resume the journey, engrave knowledge in the memory, modified as changes from one pass to another are noted. I am grateful for the impressions of Holland and Belgium from my "wild goose chase."

Pigalle Follies; I finally get to walk in the front door at the State Department. Finally, some time in Paris under less impecunious circumstances, though I'll pay the bill later. Again, glad to see Wiesbaden, and the North Atlantic from a different perspective.

A trip to the Atlantic, and One to the Pacific; Practical lessons in comparative religion in St. Louis, the Crusader's City. I had decided that I needed some grounding in the religious traditions of my indigenous ancestors. This is an example of my "Forrest Gump" level of cluelessness. I went to one of the libraries I had spent so much time in as a Junior High and High School student. Using the card catalog, a wooden cabinet filled with 3 by 5 cards arranged by the Dewey Decimal System, which served as the horse and buggy equivalent to today's Library Computer, I launched into an examination of "Indian Religion." I studied multiple volumes on Hinduism and Buddhism, and by the time I realized that Dewey had a different idea of

Indian religion, I was hooked on the panoply of belief and practice these two faiths encompassed.

Wanderlust took hold again and found me in Berkeley, Lahaina, Maui, Berkeley again, a center of people coming and going all over the world. I now had a traveling companion and was no longer the solitary pilgrim.

Serious music study mixed with yet another international cultural perspective, this one with cold war undertones.

On the Road Again; Winterreise- A winter hitch to Paris with a digression about a winter trip in '68. Back to Berkeley, from which I have not wandered far. The privilege of working a while in the City by the Bay.

Afterword

~

I used to think the Air Force, the VA or somebody owed me something for disrupting my earning power as a young man. Obviously my experience pales compared to someone who was repeatedly dosed with BZ. My story was rehearsed by summaries I sent to to various departments and agencies. Lawyers for the AF stationed at Lackland informed me of the doctrine of Sovereign Immunity, a principle evoked by Nixon when he said words to the effect of; if the President does it, it's not illegal. The CIA sent me a claim form which I duly submitted. They said basically, "fine, sue us," an impossible task. Round and round with the VA finally ended with my finally finding where my records were, requesting them, then being told I would have to supply those records to them in order to receive them. Thank you, Kurt Vonnegut.

I made peace with the idea that I would gladly wait at the back of the line until those vets and the many combat vets from the Vietnam War received their due.

The next stage was telling anyone who would sit still for it various anecdotes I thought might be interesting or instructive. I was encouraged to commit these stories to paper by Don Ellis of Grizzly Peak Press, and the rest is history.